A good grasp of the theory of structures – the theoretical basis by which the strength, stiffness and stability of a building can be understood – is fundamental to structural engineers and architects. Yet most modern structural analysis and design is carried out by computer, with the user isolated from the process in action. This book, therefore, provides a broad introduction to the mathematics behind a range of structural processes.

The basic structural equations have been known for at least 150 years, but modern plastic theory has opened up a fundamentally new way of advancing structural theory. Paradoxically, the powerful plastic theorems can be used to examine 'classic' elastic design activity, and strong mathematical relationships exist between these two approaches. Some of the techniques used in this book may be familiar to the reader, and some may be new, but each of the topics examined will give the structural engineer fresh insight into the basis of the subject.

Elements of the Theory of Structures

ELEMENTS
OF THE
THEORY OF
STRUCTURES

JACQUES HEYMAN

Emeritus Professor of Engineering
University of Cambridge

CAMBRIDGE
UNIVERSITY PRESS

CAMBRIDGE UNIVERSITY PRESS
Cambridge, New York, Melbourne, Madrid, Cape Town, Singapore, São Paulo

Cambridge University Press
The Edinburgh Building, Cambridge CB2 2RU, UK

Published in the United States of America by Cambridge University Press, New York

www.cambridge.org
Information on this title: www.cambridge.org/9780521550659

First published 1996
This digitally printed first paperback version 2006

A catalogue record for this publication is available from the British Library

Library of Congress Cataloguing in Publication data
Heyman, Jacques.
Elements of the theory of structures / Jacques Heyman.
p.   cm.
Includes bibliographical references and index.
ISBN 0 521 55065 3 (hardcover)
1. Structural analysis (Engineering)   I. Title.
TA645.H464   1995
624.1′7–dc20   95-47651   CIP

ISBN-13 978-0-521-55065-9 hardback
ISBN-10 0-521-55065-3 hardback

ISBN-13 978-0-521-03420-3 paperback
ISBN-10 0-521-03420-5 paperback

Contents

Preface

The theory of structures is one of the oldest branches of engineering. There was early interest in large, indeed ostentatious, buildings, and the design of such buildings needed more than peasant tradition; they were intended to be, and were, spectacular feats, and they required professional advice from acknowledged masters. Names of their designers are known through two or three millenia, and building manuals have survived through the same period. (Man showed also an early interest in waging war, and military engineering is another ancient profession; civil engineers are non-military engineers.)

As might be expected from an ancient discipline, the theory of structures is an especially simple branch of solid mechanics. Only three equations can be written; once they are down on paper, the engineer can in principle solve the whole range of structural problems. Sometimes the equations can be examined individually; sometimes a simple tool, virtual work, can be used to combine them to yield surprising results. In every case, however, it is only the three master equations which come into play. Equations of statics will ensure that a structure is in equilibrium. Geometrical equations will ensure that all parts of a structure fit together before and after deformation, and that the structure rests securely on its foundations. Finally, the properties of the material used to build the structure will enter the equations relating the strain in a member to the applied stress.

These equations were, effectively, known by 1826 (Navier), or more certainly by 1864 (Barré de Saint-Venant). Of course, although the equations are essentially simple, individual pieces of mathematics may become difficult. By the end of the nineteenth century, indeed, many problems had been formulated completely, but the equations were so complex that they could not usually be solved in closed form, and numerical computation was impossibly heavy. This situation gave an exhilarating spur in the twentieth century to the development of highly ingenious approximate methods of solution, and also to a fundamental reappraisal of the whole basis of the theory of structures. These developments have now been almost completely arrested

by the advent of the electronic computer; the Victorian equations, insoluble
a century ago, can now be made to yield answers. That the equations may
not be a good reflexion of reality, so that their solutions do not actually give
the required information, is only slowly being realized.

This book is concerned with the basic equations and the way in which
they should be used. The equations themselves have an intrinsic interest, as
does their application to a whole range of structural problems. The later
chapters of this book give a tiny sample, from the almost infinite number
of topics in the theory of structures, for which the results are important, or
startling, or simply amusing.

The theory of structures

A structure (from the Latin *struere*) is anything built: say an arched bridge or a cathedral from stone; a ship or a roof (and perhaps a spire) from timber; an earth dam or an excavation in soil for a fortification; or (as isolated usages) iron bars (in China first) or vegetable ropes to form suspension chains in bridges. Before the Renaissance all these structures were built without calculation, but not without 'theory', or what today would be called a 'code of practice'. Mignot's statement in 1400, at the expertise held in Milan, that *ars sine scientia nihil est* (practice is nothing without theory), testifies to the existence of a medieval rule-book for the construction of cathedrals; the few pages of a builder's manual bound in with the book of *Ezekiel* in about 600 BC show that there were yet earlier rules. These rules, for construction in the two available materials, stone and wood, were essentially rules of proportion and, as such, are effectively correct.

Stresses in ancient structures are low, and this has helped to ensure their survival. The stone in a medieval cathedral, or in the arch ring of a masonry bridge, is working at a level of one or two orders of magnitude below its crushing strength. Similarly, deflexions of such structures due to loading are negligibly small (although the movements imposed by warping of the material or by slow movements of foundations may often be seen). What is necessary is that ancient structures should be of the right form; a flying buttress must be of the right shape, an arch ring must have a certain depth, and a river pier must have a minimum width. Correct form is a matter of correct geometry, and the ancient and medieval rules of proportion were established empirically to give satisfactory designs.

The introduction two centuries ago of iron, and then steel, as a structural material generated new structural forms, such as the large-span lattice girder, or truss, and the high-rise framed building; and the advent of reinforced concrete made possible such attenuated construction as the thin shell roof. These 'modern' structures work their materials harder than earlier buildings in stone and wood; a slender steel truss will have ambient stresses which

1

are a sensible proportion of the yield stress (say one half or more). Further, higher stresses will lead to larger elastic deformations of the components of the structure; these may be small in themselves, but overall deformations may now be of importance. This higher 'performance', coupled with the increase in complexity of structural form, dictated the need for a structural theory no longer based on empirical rules of proportion – a theory which could assess whether or not a particular construction would fulfil its design criteria.

There are three main criteria which must be satisfied if a structure is to be successful (and a large number of other minor criteria that the modern designer will also take into account); they are those of strength, stiffness and stability. The homely example of a four-legged table may make clear the three aspects of performance that are being examined. The legs of the table must not break when a (normal) weight is placed on top, and the table top itself must not deflect unduly. (Both these criteria will usually be satisfied easily by the demands imposed by a dinner party.) Finally, the stability criterion may be manifest locally, or overall. If the legs of the table are slender, they may buckle when the overall load on the table is increased. Alternatively, if the legs are not at the four corners, but situated so that the top overhangs them, then placing a heavy weight near an edge may result in the whole table overturning.

1.1 The strength of beams to 1826

Galileo (1638) was concerned with the fracture strength of beams. His structure was the simple cantilever, with a value of bending moment at the root of the cantilever calculable by statics; in modern terms, he wished to evaluate the maximum value of the bending moment that could be imposed on a beam made of material having a known tensile stress at fracture. This is a problem which falls into what is now known as 'the strength of materials'; that is, it is concerned with the calculation of internal stresses in a structural member and the relation of those stresses to an observed limiting strength. Galileo deduced correctly that the strength of a beam of rectangular cross-section is proportional to the breadth and the square of the depth of the beam; he assumed that, at fracture, the whole of the cross-section was in a state of uniform tension, so that the 'neutral' axis lay in the surface of the beam. Mariotte carried out experiments in 1686, and he extended the theory by arguing that the extensions (and hence the stresses) should vary linearly with distance from the neutral axis. There were numerical mistakes in this analysis, repeated by James Bernoulli (1705), and it fell to Parent (1713) to derive the first proper account of the position of the neutral axis in elastic bending. Parent's work was not well known, and Coulomb (1773) is often credited with the solution of the elastic problem; he does indeed use explicitly a linear stress–strain relationship to derive a linearly varying stress distribution over a cross-section in flexure. However, Coulomb, like Galileo, was interested in the problem of fracture, and a full grasp of the proper basis of elastic analysis is only glimpsed in his work.

Navier gave in 1826 (with some correction and much expansion by Saint-Venant in 1864) a full account of the elastic bending of beams, that is, of the determination of the elastic stresses in a cross-section resulting from the flexure of that section by a specified bending moment. This part, and it is only a part, of the general structural problem had therefore been solved — the criterion of strength could be examined.

1.2 The stiffness and stability of beams to 1826

James Bernoulli made a start in 1691 on the determination of the shape of a bent elastic member, with his statement that the curvature at any point of an initially straight uniform beam would be proportional to the bending moment at that point. Daniel Bernoulli demonstrated in 1751 that the resulting elastic curve gave minimum strain energy in bending, and he proposed to Euler that the calculus of variations should be applied to the inverse problem of finding the shape of the curve of given length, satisfying given end-conditions of position and direction, so that the strain energy was minimized.

Euler made this analysis of the 'elastica' in 1744, and his work is summarized in chapter 6. Euler found nine separate classes of solution, and the first class, which deals with very small excursions from the linear form, is of great practical importance. Euler showed that the excursions were sinusoidal, and that they could be maintained only in the presence of a calculable value of the axial load; in practical terms, the elastica buckles in the presence of the 'Euler buckling load'.

Small lateral deflexions of a cantilever had been treated directly by Daniel Bernoulli (1751) $\left(EI\,d^2y/dx^2 = Wx\right)$, and Euler in turn made a direct study of the buckling problem in 1757 $\left(EI\,d^2y/dx^2 = -Py\right)$. Lagrange (1770) gave the first satisfactory account of higher buckling modes. All this was collected together by Navier in his *Leçons* of 1826, and he gives a recognizably modern account of the small elastic deflexions of beams.

Thus, provided that the values of the internal structural forces were known (say, the values of the bending moments from point to point of the structure), then each of the three main structural criteria (strength, stiffness, stability) could be examined with some confidence by the time of the first quarter of the nineteenth century.

1.3 Structural analysis from 1826

Of the three groups of master equations of structural analysis, the first and most fundamental is that of statical equilibrium − the internal forces in a structure (say the bending moments in a beam) must be in equilibrium with the externally applied loads. If in fact the internal forces can be found at once from the equations of statical equilibrium, then the structure is statically determinate.

Navier treated also statically indeterminate beams. He found that equations of statics alone may not be sufficient to determine the internal forces. In order to calculate these internal forces for a hyperstatic structure, the two remaining groups of equations must be used; statements of compatibility of deformation must be made, and some material laws (stress–strain relationships, effects of temperature) must be postulated. Thus, for a two-dimensional loaded beam resting on two simple supports, the reactions on those supports can be found by statics. If the same beam rests on three supports, however, only two equations are available to determine the values of three reactions. In order to proceed with the analysis, the (very small) elastic displacements of the beam must be examined, and − most importantly − boundary conditions must be specified (that the supports are rigid, for example). All this is necessary, not to compute displacements (which will be determined by the analysis, but which may or may not be of interest), but in order to solve the main structural problem, that of finding the internal forces in the (hyperstatic) structure resulting from given external loading. The stresses arising from these forces can then be calculated, and the strength criterion examined. Indeed, it was Navier who formulated clearly the elastic method of design, which is based on the actual working values of the internal forces; the resulting internal stresses should not exceed a proportion of the limiting stress for the material.

For the four-legged table, the load in each leg is to be determined for a given load on the table-top, and the corresponding stresses calculated. Now for the three-dimensional table only three appropriate equations of overall equilibirum can be written; the forces in the legs of a tripod can be found from statics, but not those in the legs of the four-legged table. To calculate these, the flexure of the table-top and the elastic compressions of the legs must be taken into account, and the problem at once becomes extremely complex. A finite-element package may be able to deal with the very large number of equations, but neither the computer nor the engineer making hand calculations will have exact knowledge of the boundary conditions; both will probably assume, unthinkingly, that the table legs are of the same length and that the floor is completely rigid.

A real table has legs of unequal length and stands on an uneven floor. Its 'actual' observable state is an accidental product of its history — it will be standing on three of its four legs, and the fourth will be unloaded. At any moment, however, a small shift of loading may rock the table (or a passing waiter may nudge it to a new position) so that the unloaded leg now carries weight, while one of the other three legs is relieved. All that can truly be said of the table is that the load in a leg lies somewhere between zero and a value calculable by simple statics. What is certain is that the sum of the forces in the *four* legs must exactly equal the weight of the table and its imposed load, but to ask for the value of the 'actual' load in an individual leg is meaningless; there are an infinite number of equilibrium states for a hyperstatic structure subjected to a given loading.

Of the three groups of master equations, those of equilibrium are the most exact. (They would appear to be perfectly exact, but they are in fact influenced marginally by the assumptions made by a structural engineer. For example, the leg of a table will be represented by a line for the purpose of calculating the load in that leg, and the precise way in which this is done affects the calculations.) The second group of equations, those defining the material properties, are less exact. The elastic moduli for steel do not show much variation, but the yield stress varies with very small changes in the chemical composition. Concrete shows a much wider variation of both elastic and strength properties, and in addition is markedly time-dependent; similarly, timber is not only anisotropic but it has variable properties from section to section.

However, it is the third group of equations, those of compatibility (and particularly the boundary conditions) which may be in many cases (as for the four-legged table) essentially unknowable. Further, just as a clearance of a fraction of a millimetre will reduce the load in a table leg to zero, so do very small shifts in the geometry of the environment alter profoundly the internal forces in a real and clearly useful and satisfactory structure. Indeed, it became clear in the 1930s, with the publications of the Reports of the Steel Structures Research Committee, that the stresses in real structures bore very little relation to those calculated confidently by the elastic designer. It was this observation which led to the development of plastic methods of design.

1.4 Plastic theory

Plastic methods are concerned with estimates of the strength of structures, and they make use of the fact that any practical material has good ductile properties. Thus glass, which would shatter, is not used to form a load-bearing structure; wrought iron or steel, however, can suffer a certain amount of permanent deformation, as can timber, reinforced concrete and, as it turns out, masonry. Such materials allow internal forces in a structure to redistribute themselves; as loads are slowly increased, their final collapse values are predictable, and reproducible, with spectacular accuracy. The small imperfections of fabrication and erection of a hyperstatic structure, which alter so markedly the elastic distribution of internal forces, have no effect on ultimate carrying capacity.

Figure 1.1 illustrates the simplest of hyperstatic structures, the uniform continuous beam on three supports. Application of the three master equations leads to the conventional elastic solution sketched in fig. 1.1(b), where the largest bending moment (of value $W\ell/8$) is found to occur at the central prop. This calculation relies, of course, on the assumption that the three supports are completely rigid, or, at least, if they do settle, then they all settle by the same amount. Any very small differential settlement, which must inevitably occur in practice, will have a large effect on the computed values of the bending moments. In particular, the largest bending moment at the central prop may be either reduced or increased, but it is on this single value that the elastic design of this uniform beam would be based.

The plastic designer imagines a hypothetical increase in loading; in fig. 1.1(a) the loads W are supposed to increase slowly. Whatever the 'actual' bending-moment distribution one value somewhere in the beam will finally exhaust the carrying capacity of the cross-section — a plastic hinge will form in a steel beam. The carrying capacity of the beam as a whole is

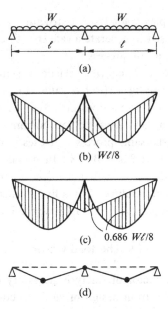

Fig. 1.1. (a) Continuous beam on three supports. (b) Conventional
elastic solution. (c) 'Plastic' solution. (d) Plastic collapse
mechanism.

not, however, exhausted at this stage; this situation is reached when a mechanism of collapse, fig. 1.1(d), finally appears. This mechanism is not at all dependent on the order of formation of the hinges, and so is not influenced by those practical imperfections which so alter the elastic solution.

The bending-moment diagram for the continuous beam which is collapsing by the mechanism of fig. 1.1(d) is shown in fig. 1.1(c). More precisely, the bending moments in fig. 1.1(c) are supposed to be due to the *working* values of the applied loads; when those values are increased in proportion by the load factor thought appropriate by the designer (say 1.75), then the diagram of fig. 1.1(c) will be stretched vertically by that same factor so that the largest bending moments just equal the full plastic moment of the section.

It must be emphasized that this increase is hypothetical. An alternative way of regarding the design process is to note that figs. 1.1 (b) and (c) both represent bending moments in the beam which are in equilibrium with the given loads. The elastic designer believes fig. 1.1(b) to be correct; the plastic designer knows that fig. 1.1(c) represents collapse. The fundamental theorem of the simple plastic analysis of structures states that either of the two diagrams may be used to generate a safe structure; a design based upon *any* equilibrium state is a safe process.

Thus, for this simple example, the 'elastic' designer will note that the largest bending moment in fig. 1.1(b) has value $W\ell/8$, and will arrange that the corresponding bending stress does not exceed a certain value, say 165 N/mm^2 for steel with a yield stress of 250 N/mm^2. The 'plastic' designer in effect goes through the same process, but uses now fig. 1.1(c) where the largest bending moment is 0.686 $W\ell/8$. In doing this, the designer has the assurance of the safe theorem that the design cannot possibly collapse until the working loads are multiplied by a hypothetical load factor of value 250/165 times the shape factor, that is 1.75 for the I-section.

1.5 The load factor

It is evident that some factor of safety is necessary in design. The factor of 1.75 arises empirically from design of steel structures in this century; a simply-supported beam sized for strength by conventional elastic codes has approximately this margin of strength. (Ancient structures of stone or wood have very much larger strength factors, but their behaviour is governed by other criteria as has been noted above.) It is not necessarily correct, however, to identify a factor such as 1.75 with a corresponding 75 per cent overload of a structure; that is, it is not necessary to imagine an actual state of collapse.

A clear definition is needed of what constitutes, and what does not constitute, a structural failure, when related to the design of that structure. Destruction of a building by bombing in time of war represents a class of loading which lies outside the normal brief of the designer (although at another time a specific design of an air-raid shelter may be needed), and such destruction is not a true structural failure. Similarly, if the ultimate strength of a particular bridge is assessed at say 25-tonne superimposed load, and as a consequence a 10-tonne load limit is placed on the bridge, then the passage of a 30-tonne load which actually causes collapse corresponds to wilful destruction rather than structural failure.

It is always possible, in fact, to imagine loading conditions which will cause collapse, but it is usually wrong to group those conditions with the 'normal' types of loading, and to pursue quasi-probabilistic calculations along the tail of some normal or skew distribution. Different types of loading do indeed have statistical distributions, but the British Standard on the structural use of concrete (for example) is clear that the load factor ('the appropriate partial safety factor') '... is introduced to take account of unconsidered possible increases in load, inaccurate assessment of load effects, unforeseen stress redistribution and variations in dimensional accuracy ...' The factors cover, therefore, a wide range of unknowable quantities, and the values given in British Standard 8110 lead to designs of straightforward structures which have much the same margin of safety as those which experience has shown to be satisfactory.

The empirical assignment of values to load factors in this way is sensible; in the absence of precise information it is right to make use of experience. But it is wrong to forget that the numerical work *has* been arranged empirically, and to come to believe that the values of load factor found to give good practical designs actually correspond to a real state of loading. The collapse mechanism of fig. 1.1(d) is a simplifying concept derived from simple plastic theory; it leads to the equilibrium bending moments for *working* (unfactored) loads sketched in fig. 1.1(c).

1.6 Deflexion and stability calculations

The structural criterion of strength may be checked from either of the equilibrium bending-moment diagrams of figs. 1.1(b) and (c). If the design is based on fig. 1.1(c) it will be more economical and will have an adequate margin of safety.

Neither bending-moment distribution, however, can be said to represent the 'actual' state of the beam, which is essentially unknowable. The 'elastic' distribution gives a mid-span deflexion of $W\ell^3/192EI$, while the 'plastic' distribution gives a corresponding deflexion which is 50 per cent greater. If the beam happens to be stiff, so that any deflexion criterion is easily satisfied, then either estimate will do. However, if the flexibility of the beam begins to govern the design, then the designer would like some guidance as to how to proceed. Such guidance cannot be given − or, rather, the designer should be aware that any estimate of deflexion may not be observable closely in the real structure. Ways of making estimates of deflexions for large and complex structures are discussed in chapter 3.

The problem is equally acute if stability calculations are pursued. For example, if the forces acting on the ends of a member of a structure are known, then that member may be checked for, say, lateral-torsional buckling.

Once again, the designer should be aware that the internal forces used for such checks may differ considerably from those actually experienced by the structure during its lifetime. A range of such equilibrium distributions should therefore be considered if the stability criterion is likely to be of significance to the design.

Virtual work

The theory of structures deals with the mechanics of slightly deformable bodies. The 'slight' deformations are such that, viewed overall, the geometry of the structure does not appear to alter, so that, for example, equilibrium equations written for the original structure remain valid when the structure is deformed. A familiar example for pin-jointed trusses arises in the resolution of forces at nodes; the inclinations of the bars are assumed to remain fixed with respect to a set of reference axes. For beams and frames, deflexions within the length of a member are 'small' compared with the length of that member, and so on.

Thus when overall equilibrium equations are written for slightly deformable structures they are identical with those obtained by rigid-body statics. The use of the equation of virtual work to obtain such relations between the external forces acting on rigid bodies can be traced back to Aristotle (384–322 BC), to Archimedes (287–212 BC) with his formulation of the laws of the lever, and, in more recent times, to Jordan of Nemore in the thirteenth century AD. The insight obtained by the use of virtual work in the study of rigid bodies is deep, but the application is straightforward. In effect, the simplifying concept of an energy balance is used to obtain relations between external forces by the study of a small rigid-body displacement (which need not be a *possible* displacement for the structure in question – hence 'virtual').

The equation of virtual work is altered profoundly when the body being studies suffers slight deformations. Not only will external forces acting on the body be involved; account must somehow be taken of the internal forces in equilibrium with those external forces. In the general application of the equation of virtual work, full use is made of the implication of the words 'in equilibrium with'; the internal forces may be any one of the (infinitely many) sets which satisfy the equilibrium equations.

If, in fact, attention is restricted to the set of elastic internal forces which are imagined to be those which actually arise from the imposition of given

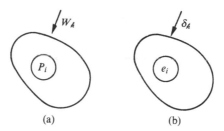

Fig. 2.1. (a) Internal forces P in equilibrium with external forces W.
(b) Internal deformations e compatible with surface
displacements δ.

loads on the structure, then the application of virtual work, although much
more complex than in the case of rigid bodies, is essentially trivial. There has
been some misunderstanding on this point. As a single example, Sokolnikoff
(1946) states the theorem of virtual work in terms of an equation between the
work done by the surface forces and the change in internal strain energy. As
such, he remarks that 'the Theorem of Virtual Work and that of Minimum
Potential Energy are seen to be merely different ways of stating the same
principle'. There seems, therefore, to be no particular virtue in retaining
a separate concept of 'virtual work', and indeed both the theorem and the
remark are omitted from Sokolnikoff's second edition of 1956. In general it
may be noted that any presentation of virtual work which involves material
properties (elastic constants and the like) is at best a highly specialized
formulation.

The essential feature of the broad statement of the principle of virtual
work is that it is not concerned with material properties, but with the other
two master equations of the theory of structures. The principle relates two
completely independent statements about a structure, namely an expression
of equilibrium of forces and an expression of compatibility of deformation.
There is no requirement that the deformations are those which arise from
the forces (if they do so arise, and are elastic, then the principle reduces to
that of Minimum Potential Energy, as noted by Sokolnikoff).

In fig. 2.1(a) a body is shown acted upon by external forces W_k, and
internal forces P_i are in equilibrium with those external forces. In fig. 2.1(b),
and as a completely separate matter, the same body is shown suffering
internal deformations e_i which lead to surface displacements δ_k. Then if
$P_i e_i$ denotes the virtual work done by the internal forces P_i moving through

displacements e_i, and similarly $W_k \delta_k$ the virtual work of the external forces W_k moving through displacements δ_k, then the principle states that

$$\Sigma W_k \delta_k = \Sigma P_i e_i. \tag{2.1}$$

In what follows the principle of virtual work will be illustrated by application to framed structures, where bending of the members is the prime structural action. Thus the internal forces P will be identified with bending moments M, and internal deformations with changes of curvature κ (and also discrete 'kinking' of a member by hinge rotations θ).

2.1 Equilibrium

The basic equilibrium equation for a straight member of a frame is

$$\frac{d^2 M}{dx^2} = w, \tag{2.2}$$

where the coordinate x is measured along the length of the member, and where w is the intensity (not necessarily uniform) of the transverse loading acting on the member. Thus the bending moments in the member are not determined uniquely by the equilibrium equation (2.2), but only as a function of two arbitrary constants of integration:

$$M = \int\int w\,dx\,dx + Ax + B. \tag{2.3}$$

The double integral in equation (2.3) is of course a particular integral of the differential equation (2.2); it corresponds to the 'free' bending moment M_w for the member. Since the whole of equation (2.3) is also a particular integral of the original equation, the free bending moments may be chosen to satisfy two numerical values for two values of x. Thus a free bending-moment diagram for the fixed-ended beam of fig. 2.2(a) may be drawn as in fig. 2.2(b), where the particular integral has been chosen to satisfy the conditions $M_w = 0$ at the two ends of the beam.

The complementary function $(Ax + B)$ in equation (2.3) is the 'reactant line' for the beam; it is sketched in fig. 2.2(c), and represents the effects of the redundant actions in the member. The complete bending-moment diagram results from the superposition of figs. 2.2(b) and (c); in effect, the bending moments in fig. 2.2(d) are measured from a sloping base line. The theory of structures has as its primary objective the establishment of the position of

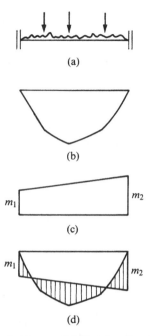

Fig. 2.2. (a) Fixed-ended beam. (b) Free bending-moment diagram.
(c) Reactant line. (d) General bending-moment diagram.

this base line, that is, with the determination of the values of the redundant quantities.

When individual members are assembled together to give the general framed structure, the bending moment at any section may be written

$$M = M_w + \sum_{r=1}^{N} \alpha_r M_r \tag{2.4}$$

where the redundant quantities M_r arise from equilibrium equations such as equation (2.3); the number N of these redundancies depends on the way the members are connected to each other and to external supports. The functions α_r are linear functions of the coordinate x which is used to define the location of the sections of the frame, so that equation (2.4) gives

$$\frac{d^2 M}{dx^2} = \frac{d^2 M_w}{dx^2} (= w), \tag{2.5}$$

and the fundamental equilibrium equation is satisfied by any particular set of

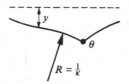

Fig. 2.3. Originally straight member deformed with elastic curvatures κ
and hinge discontinuities θ.

free bending moments on which is superimposed an arbitrary set of bending
moments arising from the redundant quantities M_r. In words, equations (2.4)
and (2.5) state that:

$$\left.\begin{array}{l}\text{Bending moments } M \text{ are in equilibrium with external loads } w.\\ \text{Bending moments } M_w \text{ are in equilibrium with external loads } w.\\ \text{Bending moments } \alpha_r M_r \text{ are in equilibrium with zero external loads.}\end{array}\right\} \quad (2.6)$$

The last of these three statements implies that the bending moments $\alpha_r M_r$
are self-stressing (or residual) moments; these cannot arise in statically
determinate structures.

2.2 Compatibility of deformations

An originally straight member is deformed so that it bends continuously
with a varying curvature κ (fig. 2.3); in addition, hinge discontinuities θ may
be present. These deformations cause lateral deflexions y of the member.
Then, with the usual assumption of small deformations,

$$\kappa = \frac{d^2 y}{dx^2}, \qquad (2.7)$$

and the deflexions y are said to be compatible with the hinge discontinuities
θ and the curvatures κ.

2.3 The equation of virtual work

Equation (2.1) may be rewritten for members whose prime structural action
is that of bending in the form

$$\Sigma W_k y_k + \oint wy\,dx = \Sigma M_i \theta_i + \oint M\kappa\,dx. \qquad (2.8)$$

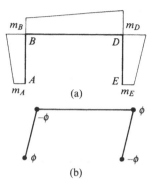

Fig. 2.4. (a) Reactant lines for fixed-base portal frame. (b) Virtual mechanism of collapse.

On the left-hand side the summation includes all point loads W, and the integral extends over all the distributed loads; on the right-hand side the summation includes all hinge discontinuities θ, and the integral extends over the rest of the frame. Equation (2.8) may be used to derive techniques of elastic and plastic methods of analysis of framed structures. It may be repeated that there is no necessary interdependence of the equilibrium statement $(W, w; M)$ and the compatibility statement $(y; \theta, \kappa)$.

2.4 Self-stressing moments

Self-stressing moments $m (\equiv \Sigma \alpha_r M_r)$ are, by definition, in equilibrium with zero external load. If an unloaded structure is given an arbitrary set (y, θ, κ) of compatible deformations, then equation (2.8) states that

$$\Sigma m_i \theta_i + \oint m \kappa dx = 0. \qquad (2.9)$$

The most general state of self-stress of the fixed-base portal frame sketched in fig. 2.4(a) is indicated by the bending-moment diagram shown. Reactant lines are known to be straight, so that if residual moments m_A, m_B, m_D and m_E are defined at the corners of the frame, the whole bending-moment diagram may be drawn. The question arises as to whether the values of the four residual moments can be assigned arbitrarily, or whether there might exist relationships between their values.

Figure 2.4(b) shows a virtual deformation of the frame in which deflexions arise solely from hinge discontinuities at the ends of the members, the portions between hinges remaining straight with $\kappa = 0$. Equation (2.9) thus reduces to the single summation term, and when it is applied using the equilibrium and deformation states shown respectively in fig. 2.4(a) and (b), it leads at once to

$$m_A - m_B + m_D - m_E = 0. \tag{2.10}$$

Thus only three of the four residual moments can have arbitrary values; the fourth is then calculable from equation (2.10). (It is of course known from general structural theory that the fixed-base arch form, such as a portal frame, has three redundancies, and these may be regarded as any three of the four moments m marked in fig. 2.4(a).)

2.5 Plastic collapse under proportional loading

The loads W acting on a given frame (having known values M_p of full plastic moment) may be imagined to be increased slowly in proportion. Simple plastic collapse will occur when enough plastic hinges have formed to turn the structure (or part of the structure) into a mechanism of one (or by accident more than one) degree of freedom. Under the action of the working value of the loads, an equilibrium distribution of bending moment in the frame may be written (cf. equation (2.4)):

$$M = M_w + m. \tag{2.11}$$

When all loads have been increased in the same ratio λ, a corresponding set of bending moments in equilibrium with these enhanced loads is clearly

$$\lambda M = \lambda M_w + \lambda m. \tag{2.12}$$

The value of λ at collapse, λ_c, is sought.

At collapse, in a mechanism with hinge rotations θ, the bending moments at the hinge positions will have values (M_p), or, more precisely, $(M_p)_i$, since the values may be different at different positions. A mechanism of collapse of a simple portal frame is shown schematically in fig. 2.5(a), where the hinge rotations (which permit motion of the mechanism) exist in the presence of some elastic deformations of the members. For the purpose of analysis, this actual mechanism is replaced by the 'virtual' mechanism of fig. 2.5(b), in which the *only* deformation in the frame arises from the hinge rotations, the members otherwise remaining straight. Thus, for the collapse loading, the

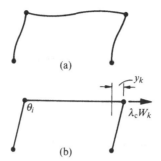

Fig. 2.5. (a) Actual mechanism of collapse. (b) Virtual mechanism.

equation of virtual work (2.8) may be written

$$\lambda_C \left[\Sigma W_k y_k + \oint wy\,dx \right] = \Sigma M_i \theta_i. \qquad (2.13)$$

Now the values of M_i for the collapse mechanism of fig. 2.5(a) will have, as noted, the values $(M_p)_i$ at each hinge location. Moreover, the sign of a hinge rotation will accord with the sign of the bending moment at that hinge, so that equation (2.13) may be written

$$\lambda_C \left[\Sigma W_k y_k + \oint wy\,dx \right] = \Sigma (M_p)_i |\theta_i|. \qquad (2.14)$$

A second application of the equation of virtual work may be made using the general equilibrium distribution of equation (2.12), that is

$$\lambda_C \left[\Sigma W_k y_k + \oint wy\,dx \right] = \lambda_C \left[\Sigma (M_w)_i \theta_i + \Sigma m_i \theta_i \right]. \qquad (2.15)$$

The term $\Sigma m_i \theta_i$ in equation (2.15) is of course zero, from equation (2.9), so that equations (2.14) and (2.15) combine to give

$$\lambda_C \Sigma (M_w)_i \theta_i = \Sigma (M_p)_i |\theta_i|. \qquad (2.16)$$

Equation (2.16) is the basic collapse equation from which the collapse load factor λ_C can be calculated for a known collapse mechanism θ. If the collapse mechanism θ is assumed to be correct, but not actually known to be so, then it follows easily from further applications of the equation of virtual work that the resulting value of the load factor λ is an upper bound on the true value at collapse, that is $\lambda \geq \lambda_C$.

The bending moments M_w in the collapse equation (2.16) represent *any* set of equilibrium moments. Thus the fixed-ended beam of fig. 2.6(a) has the

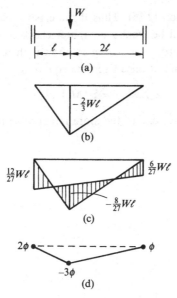

Fig. 2.6. (a) Fixed-ended beam. (b) Corresponding free bending moments. (c) Elastic solution. (d) Plastic collapse mechanism.

'free' bending moments of fig. 2.6(b), and these may be used as the set M_w. The 'virtual' collapse mechanism is shown in fig. 2.6(d), so that

$$\lambda_c \left(-\frac{2}{3} W\ell \right) (-3\phi) = M_p (2\phi + 3\phi + \phi),$$

$$\text{or } \lambda_c = \frac{3M_p}{W\ell}. \tag{2.17}$$

Another equilibrium distribution M_w is shown in fig. 2.6(c), in which the reactant line has been positioned to give the bending moments shown. Equation (2.16) may be written

$$\lambda_c \left[\left(\frac{12}{27} W\ell \right) (2\phi) + \left(-\frac{8}{27} W\ell \right) (-3\phi) + \left(\frac{6}{27} W\ell \right) (\phi) \right] = 6M_p\phi, \tag{2.18}$$

which reduces at once to equation (2.17).

The reactant line has not of course been positioned arbitrarily in fig. 2.6(c), although in fact equation (2.17) will be recovered for any arbitrary placing of the line. Figure 2.6(c) represents the elastic solution for the beam under the given loading; elastic moments, $\mathcal{M}$ say, are by definition in equilibrium with the external loading, and hence may be used as unique and particular values of M_w in equation (2.16). Thus if the elastic solution for a frame is available, as it may well be from a computer check of deflexions in a final design, and if the collapse mechanism is known, then the plastic collapse load factor may be found at once from the equation

$$\lambda_c \Sigma \mathcal{M}_i \theta_i = \Sigma \left(M_p \right)_i |\theta_i| . \tag{2.19}$$

Equation (2.19) illustrates one of the more remarkable links between elastic and plastic analysis.

Betti, Maxwell,
Müller-Breslau, Melchers

The equation of virtual work will be used to derive some of the familiar elastic theorems. Equation (2.8) may be simplified without loss of generality by supposing that all external loads are point loads; further, since elastic deformations only are under consideration, there will be no hinge discontinuities. For a framed structure, then, the equation becomes

$$\Sigma W_k y_k = \oint M \kappa dx, \tag{3.1}$$

where internal bending moments M are in equilibrium with external loads W_k, and where internal curvatures κ of the members lead to (are compatible with) displacements y_k at the points of application of the loads. As usual, there is no necessary connection between the equilibrium statement (W_k, M) and the compatibility statement (y_k, κ).

Thus in fig. 3.1(a) the external loads W_k^* are in equilibrium with internal bending moments M^*, while, as a completely separate matter, fig. 3.1(b) indicates a possible pattern of deformation of the same frame. Then equation (3.1) may be used to relate the two pictorial statements of fig. 3.1:

$$\Sigma W_k^* y_k = \oint M^* \kappa dx \tag{3.2}$$

Now if the bending moments M^* give rise to elastic curvatures κ^*, then

$$M^* = EI\kappa^*, \tag{3.3}$$

and equation (3.2) may be written

$$\Sigma W_k^* y_k = \oint EI\kappa^* \kappa dx, \tag{3.4}$$

where EI represents the flexural rigidity of the members of the frame.

The actual curvatures κ^* and corresponding deflexions y_k^* are associated

21

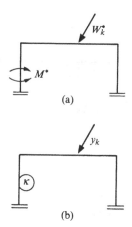

Fig. 3.1. (a) Bending moments M^* in equilibrium with external loads
W_k^*. (b) Curvatures κ compatible with displacements y.

with the equilibrium system of fig. 3.1(a), and similarly loads W_k and associated bending moments M may be imagined to lead to the deformations sketched in fig. 3.1(b). If therefore the roles of figs. 3.1(a) and (b) are interchanged, then equation (3.4) is replaced by

$$\Sigma W_k y_k^* = \oint EI\kappa\kappa^* dx. \tag{3.5}$$

3.1 Betti's reciprocal theorem, 1872

When equations (3.4) and (3.5) are compared, it is evident that

$$\Sigma W_k^* y_k = \Sigma W_k y_k^*, \tag{3.6}$$

and this is Betti's reciprocal theorem. In words, if two actual states (starred and unstarred) of an elastic body are considered, then the work done by the loads W_k^* of the first state on the displacements y_k of the second state is equal to the corresponding work done by W_k on y_k^*.

That Betti's theorem applies to statically indeterminate frames is clear from its derivation. The addition of a residual moment m^* to the equilibrium moments M^* would expand equation (3.2) to

$$\Sigma W_k^* y_k = \oint M^*\kappa dx + \oint m^*\kappa dx, \tag{3.7}$$

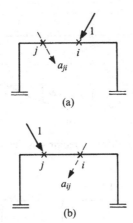

Fig. 3.2. Maxwell's reciprocal theorem: (a) unit load at i produces
elastic deflexion a_{ji} at j, and (b) unit load at j produces
elastic deflexion a_{ij} at i. Then $a_{ij} = a_{ji}$.

and the final integral is zero by virtue of the fact that the moments m^* are
in equilibrium with zero external load. Thus the arguments are the same if
M^* are the actual elastic bending moments or simply bending moments in
equilibrium with the external loads W_k^*.

3.2 Maxwell's reciprocal theorem, 1864

Maxwell's reciprocal theorem for linear systems is effectively a statement of
the symmetry of the matrix of elastic flexibility coefficients. In fig. 3.2(a) a
unit load applied in a specified direction at section i of a frame produces
an elastic deflexion a_{ji} in a specified direction at section j of the frame.
Similarly, an elastic deflexion a_{ij} results at section i from the application
of a unit load at section j (fig. 3.2(b)). Inserting these two statements into
equation (3.6) gives at once

$$a_{ij} = a_{ji}. \tag{3.8}$$

3.3 Müller-Breslau 1883,1884: Influence lines

According to Charlton, the reciprocal properties of linear statically determinate structures were known to Mohr. Specifically, Mohr considered the
simply supported beam of fig. 3.3(a), and he showed that if a load P applied

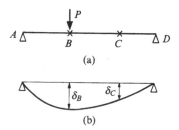

(a)

(b)

Fig. 3.3. Mohr's reciprocal theorem.

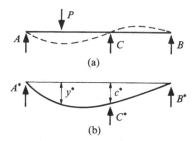

(a)

(b)

Fig. 3.4. Müller-Breslau's discussion of a statically indeterminate beam.

at B produced a deflexion δ_C at C, fig. 3.3(b), then the same load P applied at C would produce the same deflexion δ_C at B. This is, of course, a special statically determinate application of equation (3.8). Moreover, he concluded that the shape of the deflected beam sketched in fig. 3.3(b), which results from the application of the load P at section B, is in fact the same as the influence line for the deflexion at B as the load P crosses the beam.

Müller-Breslau extended these ideas to the statically indeterminate structure. In fig. 3.4(a) is illustrated his own example of a simply supported beam with an additional internal support, the system thus being once statically indeterminate. It is required to find, say, the reaction C due to the external load P. In fig. 3.4(b) a small displacement c^* of the supposedly rigid support at C has been imposed on the otherwise unloaded beam, inducing reactions A^*, B^* and C^* at the three supports. If equation (3.6) is applied to the two states sketched in fig. 3.4, then

$$(A)(0) + (P)(y^*) + (C)(-c^*) + (B)(0)$$
$$= (A^*)(0) + (C^*)(0) + (B^*)(0), \qquad (3.9)$$

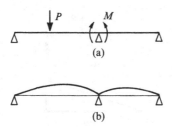

(a)

(b)

Fig. 3.5. Extension of Müller-Breslau's principle.

that is,

$$C = \frac{y^*}{c^*}P \qquad (3.10)$$

Thus, for an arbitrary unit displacement $c^* = 1$ of the internal support, and for a unit load P, the value of the reaction C at the internal support is equal to y^*. Once again fig. 3.4(b) gives, to some scale, the influence line for the statically indeterminate reaction C as a unit load P crosses the bridge.

This is Müller-Breslau's principle, and it may be used to determine internal forces in a frame. In fig. 3.5, for example, an imaginary arbitrary (unit) 'kink' has been introduced at the internal support in order to determine the bending moment M at that support which results from the application of the load P. Once again, the deflected form of the beam in fig. 3.5(b) gives the influence line; that is, fig. 3.5(b) is a plot to some scale of the value of the bending moment M as the (unit) load P crosses the beam.

It may be noted that imaginary internal discontinuities of this sort will produce linear influence lines for statically determinate structures. Such structures cannot sustain self-stress, and small imperfections, of manufacture or assembly for example, do not introduce internal forces. In fig. 3.6(a) the simply supported beam is subjected to a point load P. The influence line for the reaction at support B is found by moving the support through a unit distance; see fig. 3.6(b). The influence line for the shear force at X is found by introducing a unit lateral discontinuity (and no discontinuity in slope); see fig. 3.6(c).

3.4 Beggs 1927

The right-hand side of equation (3.9) is zero by virtue of the fact that the supports in the original real beam of fig. 3.4 are rigid. Indeed, any system of

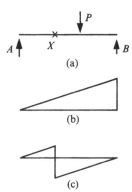

Fig. 3.6. (a) Simply supported beam. (b) Influence line for reaction at
support B. (c) Influence line for shear force at X.

deformation similar to that of fig. 3.4(b), for which imaginary displacements
are introduced at supports which are in fact rigid, or similar to that of
figs 3.5(b) or 3.6(c), in which an imaginary internal dislocation is imposed,
will lead to an equation of the form

$$\Sigma W_k y_k^* = 0 \qquad\qquad (3.11)$$

(cf. equations (3.6) and (3.9)). Since equation (3.11) is homogeneous in the
starred deflexion components, it would be possible to make the imaginary
displacements, not on the real structure, but on a scale model having the
same flexural characteristics as the real structure. All that is required is that
the scale model should have flexural rigidities that are the same constant
proportion from section to section as those of the original.

Beggs proposed that, instead of imaginary deformations, real deformations
should be imposed on a carefully made and properly scaled celluloid model.
Such a model can be cut from a plastic sheet of uniform thickness, the
depths of the members being varied to ensure correct values of the flexural
rigidities. The required coefficients (e.g. y^* and c^* in equation (3.10)) can then
be obtained experimentally. Observations of this kind can be very accurate,
and it is possible to obtain acceptable estimates if the frame is cut from
cardboard and tested pinned to a drawing board.

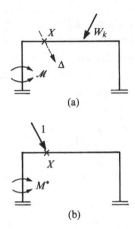

Fig. 3.7. (a) Loads W_k produce elastic bending moments $\mathcal{M}$ and a
deflexion Δ at X. (b) Bending moments M^* in equilibrium
with unit load at X.

3.5 The unit dummy-load method: Melchers 1980

The equation of virtual work may be used to calculate the elastic deflexions
of structures by means of the device of a unit dummy load. In fig. 3.7(a) loads
W_k are shown acting on a frame, and it is required to calculate the resulting
elastic deflexion Δ at a certain location X. The elastic bending moments
in the frame are denoted $\mathcal{M}$. In fig. 3.7(b) the only load on the frame is
one of unit intensity at X, acting in the direction in which the deflexion Δ
is measured. This unit load is in equilibrium with bending moments M^*,
where, as usual, the set M^* is *any* distribution satisfying the requirements of
equilibrium. Then, with fig. 3.7(a) representing the compatible displacement
set and fig. 3.7(b) the equilibrium set, the virtual work equation (3.1) states
that

$$1.\Delta = \oint M^* \frac{\mathcal{M}}{EI} dx. \qquad (3.12)$$

The device of the unit load has been introduced to give a single term,
the required deflexion, on the left-hand side of the equation. This is the
conventional way of using virtual work for deflexion calculations; the data
required are the actual elastic solution under the real loading, and any
equilibrium solution for the unit dummy load.

Melchers proposed in 1980 an inversion of this process. First, it may be

noted that the elastic flexural rigidity EI in equation (3.12) could be attached
to M^* rather than to $\mathcal{M}$. Second, the elastic moments $\mathcal{M}$ must be expressible
as any equilibrium distribution M_w on which is superimposed an appropriate
self-stressing distribution m. Thus equation (3.12) may be written

$$
\begin{aligned}
\Delta &= \oint \frac{M^*}{EI} \mathcal{M} dx \\
&= \oint \frac{M^*}{EI} (M_w + m) dx \\
&= \oint \frac{M^*}{EI} M_w dx + \oint \frac{M^*}{EI} m dx.
\end{aligned}
\tag{3.13}
$$

Now in the second integral of equation (3.13), if the curvatures M^*/EI
were elastic and the resulting displacements satisfied the real boundary
conditions (e.g. zero slope at a fixed end, and so on), then they would form
a compatible set of deformations; since the moments m are in equilibrium
with zero external load, then in that case the second integral would be zero.
Denoting, then, the *elastic* moments in equilibrium with the unit dummy
load by $\mathcal{M}^*$,

$$
\Delta = \oint M_w \frac{\mathcal{M}^*}{EI} dx,
\tag{3.14}
$$

and this is an alternative to equation (3.12), where the data required are now
the elastic solution for the unit dummy load and any equilibrium solution
for the actual loading.

Equation (3.14) is of some help in estimating working-load deflexions of a
frame that has been designed plastically against collapse. The set M_w could
for example be the collapse bending moments (under a unit load factor
rather than the collapse load factor); the elastic bending moments $\mathcal{M}^*$ for
the unit load must be determined. Thus the central deflexion of a fixed-ended
beam carrying a uniformly distributed load, fig. 3.8(a), may be found *either*
from the distributions of figs. 3.8(b) and (c), using equation (3.12), *or* from
the distributions of figs. 3.8(d) and (e), using equation (3.14).

3.6 Approximate calculations for deflexions

Exact calculations of deflexions of elastic structures demand, of course, the
full solution of the structural equations. The calculations can be sequential
for a statically determinate structure; the equations of equilibrium may be
solved by themselves to give the values of the internal forces which result

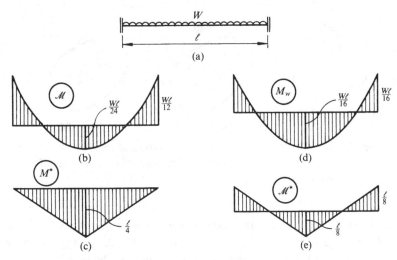

Fig. 3.8. (a) Fixed-ended beam. A uniformly distributed load produces
(b) elastic bending moments or (d) collapse bending
moments. A central point load produces (c) 'free' bending
moments or (e) elastic bending moments.

from the application of the external loads; the equations expressing the material properties will then give the values of the consequential internal deformations; and finally geometrical considerations will transform those internal deformations into external displacements of the structure.

Such a sequence cannot be followed for a hyperstatic structure. The calculations require the simultaneous solution of all three groups of the master structural equations; the work is much heavier, and considerable attention has been given in recent years to the derivation of economical computational schemes. It is of interest that approximate calculations, which abandon the attempt to solve all the equations simultaneously, can lead to meaningful estimates of deflexions. Thus the calculations are much simplified if equations of compatibility are put on one side, and attention concentrated on the equilibrium equations; similarly, deformation patterns can be assumed, without attempting to satisfy the equilibrium equations.

Thus, in fig. 3.9(a) an elastic body is acted on by a single load W, and the deflexion δ at the point of application is required. The actual bending moments in the body (taken to be a framed structure for this discussion) are $\mathcal{M}$, and the corresponding elastic curvatures are κ. Figure 3.9(b) indicates a

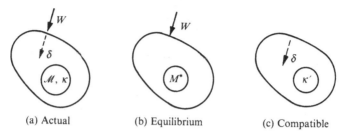

(a) Actual   (b) Equilibrium   (c) Compatible

Fig. 3.9. (a) A load W acts on an elastic body, and produces a
deflexion δ. (b) A possible equilibrium state. (c) A state of
deformation which gives the same deflexion δ.

possible equilibrium state of the body, in which bending moments M^* satisfy
the statical equations involving the external load W. An energy balance is
used to give an estimate Δ of the actual deflexion δ, in the form

$$\frac{1}{2}W\Delta = \frac{1}{2}\oint \frac{(M^*)^2}{EI}dx,$$

or, dropping the factors of $\frac{1}{2}$,

$$W\Delta = \oint \frac{(M^*)^2}{EI}dx. \tag{3.15}$$

Then

$$\Delta \geq \delta. \tag{3.16}$$

The proof of inequality (3.16) follows at once from the fact that the equilibrium moments M^* differ from the elastic moments $\mathcal{M}$ by a set of self-stressing
moments m, that is, $M^* = \mathcal{M} + m$. Thus equation (3.15) becomes

$$W\Delta = \oint \frac{(\mathcal{M}+m)^2}{EI}dx,$$

$$\text{or} \quad W\Delta = \oint \frac{\mathcal{M}^2}{EI}dx + 2\oint m\frac{\mathcal{M}}{EI}dx + \oint \frac{m^2}{EI}dx. \tag{3.17}$$

In this expression, the first integral involving the actual moments $\mathcal{M}$ may
be replaced by $W\delta$, while the second integral is zero (as usual) since the
moments m are in equilibrium with zero external loads and the curvatures

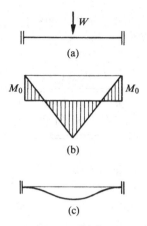

Fig. 3.10. (a) A fixed-ended beam. (b) Its general bending-moment
diagram. (c) A reasonable deformation of the loaded beam.

$\kappa = \mathcal{M}/EI$ are a compatible set. Thus

$$W\Delta = W\delta + \oint \frac{m^2}{EI}dx, \tag{3.18}$$

and the remaining integral is, of course, positive definite; hence inequality
(3.16).

As a very simple example, the fixed-ended beam of fig. 3.10 has the general
equilibrium bending-moment distribution sketched in fig. 3.10(b), which, for
$x \leq \dfrac{\ell}{2}$, is given by

$$M^* = \left(M_0 - \frac{Wx}{2} \right). \tag{3.19}$$

Equation (3.15) gives

$$W\Delta = \frac{2}{EI} \int_0^{\frac{\ell}{2}} \left(M_0 - \frac{Wx}{2} \right)^2 dx$$

$$\text{or} \quad \Delta = \frac{W\ell^3}{192EI} \left[4 - 6\left(\frac{8M_0}{W\ell} \right) + 3\left(\frac{8M_0}{W\ell} \right)^2 \right]. \tag{3.20}$$

Any value of M_0 substituted into equation (3.20) will give a value Δ greater
than the true value; indeed, the term in square brackets is a minimum for
$M_0 = W\ell/8$, which is the correct elastic solution to the problem. The

following table shows that estimates of Δ are not very good unless a fairly good estimate of M_0 is made.

Table 3.1.

$8M_0/W\ell$	0	0.2	0.4	0.6	0.8	1.0	1.2	1.4	etc
Δ/δ	4	2.92	2.08	1.48	1.12	1	1.12	1.48	

By contrast, a reasonable deformation pattern κ' will usually lead to a good estimate of the actual deflexion δ. For example, the sketched deflexions of fig. 3.10(c) — which are in fact a cubic function — may be represented by

$$y = \frac{1}{2}\Delta'\left(1 - \cos\frac{2\pi x}{\ell}\right),\tag{3.21}$$

so that

$$\kappa' = \Delta'.\frac{2\pi^2}{\ell^2}\cos\frac{2\pi x}{\ell}.\tag{3.22}$$

An energy balance in the form

$$\frac{1}{2}W\Delta' = \frac{1}{2}\oint EI\,(\kappa')^2\,dx$$

$$\text{or}\quad W\Delta' = \oint EI\,(\kappa')^2\,dx\tag{3.23}$$

may be used again to calculate the estimate Δ'. Introducing equation (3.22) into equation (3.23),

$$W\Delta' = EI\Delta'^2\int_0^\ell \frac{4\pi^4}{\ell^4}\cos^2\frac{2\pi x}{\ell}\,dx,\tag{3.24}$$

from which

$$\Delta' = \frac{W\ell^3}{2\pi^4 EI} = \frac{1}{194.8}\frac{W\ell^3}{EI}.\tag{3.25}$$

The value 194.8 is close to the correct value 192; moreover, the value of Δ' is less than the correct value δ. Indeed, calculations made in this way are subject to the inequality

$$\Delta' \leq \delta.\tag{3.26}$$

The proof of inequality (3.26) follows from the fact that the strain energy associated with arbitrary deformations κ' is always greater than that associated with the actual deformations κ. In fig. 3.9(c) an arbitrary but possible set of deformations κ' has been imposed on the body which is compatible with the *actual* deflexion δ of the loaded point in fig. 3.9(a). The equilibrium set $(W, \mathcal{M})$ of fig. 3.9(a) may be used in the equation of virtual work first with the compatible set (δ, κ) of fig. 3.9(a) and second with the set (δ, κ') of fig. 3.9(c), that is

$$\left. \begin{array}{rcl} W\delta & = & \oint \mathcal{M}\kappa dx \\ \text{and} \quad W\delta & = & \oint \mathcal{M}\kappa' dx, \end{array} \right\}$$

$$\text{or} \quad \oint \mathcal{M}\left(\kappa' - \kappa\right) dx = 0,$$

$$\text{i.e.} \quad \oint EI\kappa\left(\kappa' - \kappa\right) dx = 0. \tag{3.27}$$

Now (twice) the difference between the strain energies of the two states is given by

$$\oint EI\left\{(\kappa')^2 - (\kappa)^2\right\} dx$$

$$= \oint EI\left(\kappa' - \kappa\right)^2 dx + 2\oint EI\kappa\left(\kappa' - \kappa\right) dx, \tag{3.28}$$

and, on the right-hand side, the first integral is positive definite, and the second is zero.

Figure 3.10(c) is sketched for the actual displacement δ of the loading point; the actual calculation will be made with a factor a (cf. the factor Δ' in equation (3.24)) in the form

$$Wa = a^2 \oint EI\left(\kappa'\right)^2 dx; \tag{3.29}$$

since the value of strain energy associated with κ' is greater than the actual value, by equation (3.28), so the value of a given by equation (3.29) will be a maximum for the actual curvatures κ; hence inequality (3.26).

Jettied construction

A standard undergraduate problem in the theory of statically determinate structures concerns the 'optimum' way to lift a uniform beam. In fig. 4.1 a beam of length ℓ is suspended by two slings S situated symmetrically at distances $\alpha\ell$ from the ends of the beam. The corresponding bending-moment diagram is sketched in fig. 4.1(b), and the values of the two bending moments indicated are

$$\left.\begin{aligned} M_S &= (2\alpha)^2 \frac{W\ell}{8}, \\ M_C &= (1 - 4\alpha) \frac{W\ell}{8}. \end{aligned}\right\} \tag{4.1}$$

Evidently an increase in the value of α will increase the value of the bending moment at the points of support and decrease that at mid-span; the optimum arrangement of the slings will make the values of M_S and M_C equal. Using

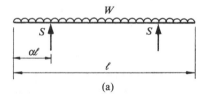

(a)

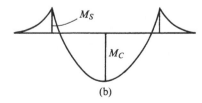

(b)

Fig. 4.1. A beam on two supports.

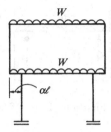

Fig. 4.2. A two-storey timber-framed house.

the two values in equations (4.1),

$$4\alpha^2 + 4\alpha - 1 = 0$$

$$\text{or} \qquad \alpha = \frac{1}{2}\left(\sqrt{2} - 1\right) = 0.2071, \qquad (4.2)$$

and, for this value of α,

$$M_S = M_C = (0.1716)\,\frac{W\ell}{8}. \qquad (4.3)$$

It may be noted that, had the beam been lifted from its ends ($\alpha = 0$), the largest bending moment would have had value $W\ell/8$; the optimum lift reduces bending stresses substantially.

4.1 A jettied house

In medieval timber-framed construction of small houses, the first storey often jetties out over the ground floor walls. A two-storey house is shown schematically in fig. 4.2, and, for the sake of simple calculations, the loads on the roof beams and the first-floor beams are supposed to have the same value. Thus, in fig. 4.3, the first-floor beam carries not only its own (uniformly distributed) weight W, but also end loads of $\frac{1}{2}W$. The bending-moment diagram for the beam has the same general features of that shown in fig. 4.1(b), and the cardinal values of the bending moments are

$$\left.\begin{aligned} M_S &= \left(4\alpha^2 + 4\alpha\right)\left(\frac{W\ell}{8}\right), \\ M_C &= (1 - 8\alpha)\left(\frac{W\ell}{8}\right). \end{aligned}\right\} \qquad (4.4)$$

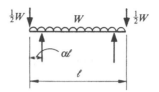

Fig. 4.3. First-floor beam of fig. 4.2.

The optimum amount of jetty may be determined by equating expressions
(4.4), from which

$$4\alpha^2 + 12\alpha - 1 = 0$$

$$\text{or} \qquad \alpha = \frac{1}{2}\left(\sqrt{10} - 3\right) = 0.0811. \qquad (4.5)$$

For this value of α,

$$M_S = M_C = (0.3509)\frac{W\ell}{8}. \qquad (4.6)$$

The best value for α will of course depend on the actual proportion of
roof load to floor load in a house; nevertheless the idealized calculations for
the equal loads of fig. 4.2 show clearly that a very small amount of jetty
reduces bending stresses dramatically. From equation (4.5) a house of width
6 m at first-floor level should oversail the ground floor by about $\frac{1}{2}$ m, a not
untypical value, and the largest bending moments in the first-floor beams
will then be reduced to about one-third of the unjettied values.

4.2 Beams with knee braces

Designs for statically indeterminate beams cannot always be 'optimized' in
the simple way given above. Knee braces were often used in timber construction to help support long-span beams, and, as for jettied construction, their
use can lead to a great reduction in stresses. Figure 4.4 shows the fifteenth-
century roof construction of the Great Dormitory at Durham Cathedral; the
main beams span over 12 m. Figure 4.5 shows some of the eleventh-century
timbers forming the belfry floor of St Albans Abbey; massive diagonal props
support the main beams, which span over 9 m. In either case the essential
load bearing member may be represented as the beam on four supports
sketched in fig. 4.6.

There are several assumptions implicit in the sketch of fig. 4.6. The loading
has been taken as uniform, which may be roughly true for St Albans; at

Fig. 4.4. Durham Cathedral: Great Dormitory.

Durham, however, the beam is acted upon by more or less point loads transmitted by the purlins. The walls provide support forces R, and these are shown as equal in fig. 4.6, whereas it is unlikely that this assumption can be justified by symmetry of the actual structure. In the same way the braces are represented in fig. 4.6 by two equal support forces S. There are other equally important and essentially unjustifiable assumptions which are noted below.

However, a first analysis will be made on the basis of the sketch of fig. 4.6, in which the extra supports S are placed at a distance $\alpha\ell$ from the walls. By analogy with the 'undergraduate problem' of the statically determinate beam, the effects of varying the value of α will be studied, and an attempt made to optimize the placing of the knee braces.

It may be imagined that, for small values of α (shorter braces than those

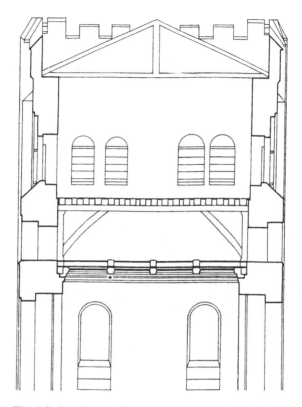

Fig. 4.5. St Albans Abbey: construction of belfry floor.

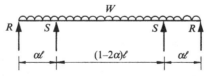

Fig. 4.6. Beam on four supports.

shown in figs. 4.4 and 4.5), the ends of the beam of fig. 4.6 will lift off the walls, so that each beam is supported only by the two forces S; the situation is that of fig. 4.1, with no reaction being supplied by the walls.

(a) No support from the walls; $\alpha < 0.2142$

It will be seen from the next section that, within the framework of assumptions, no support is provided from the walls if α is less than 0.2142. Equations

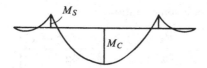

Fig. 4.7. Bending-moment diagram for fig. 4.6.

(4.1) apply, and the optimum value of α from equation (4.2), 0.2071, makes M_S and M_C equal (fig. 4.1(b)). As α is increased further the value of M_S increases above the relatively least value of equation (4.3); however, for α greater than 0.2142 (as will be seen) the walls provide support, and a different structural regime must be analysed.

(b) Support from the walls; $\alpha > 0.2142$

The general bending-moment diagram for the statically indeterminate beam of fig. 4.6 is sketched in fig. 4.7. A conventional elastic analysis will be made, and a further range of more or less doubtful assumptions will be built into the calculations. For example, unless the working is to become excessively tedious, the beam must be taken to have uniform section throughout its length, and the basic elasticity of the wood will also be taken as uniform. Both of these assumptions are clearly only very rough representations of eleventh- and fifteenth-century timbers.

Further, it is as usual the boundary conditions which influence critically the results of any attempted elastic analysis; some decision must be made, in order to proceed with the calculations at all, about the possible displacements of the four support points R and S in fig. 4.6. It has been noted already that symmetry has been assumed; any 'usual' elastic analysis (a computer package, say) will assume further, and without remark, that all four supports are rigid. In fact, any small differential settlement of the supports will have a marked effect on the values of the bending moments in the beam.

However, no assumption other than rigidity of the supports is really possible, since the support conditions are, in essence, unknowable. They depend on such factors as compressibility of the raking braces, shrinkage of connexions between the timbers, possible decay of the wall plates, and so on. The following values, which result from a 'usual' elastic analysis, must therefore be viewed with some caution:

$$R = \frac{W\ell}{4\alpha}\left[\frac{-1+6\alpha-6\alpha^2-\alpha^3}{3-4\alpha}\right],$$

$$M_S = \frac{W\ell}{8}\cdot 2\left[\frac{1-6\alpha+12\alpha^2-7\alpha^3}{3-4\alpha}\right], \qquad (4.7)$$

$$M_C = \frac{W\ell}{8}\left[\frac{1-4\alpha+4\alpha^2-2\alpha^3}{3-4\alpha}\right].$$

The value of R must be positive; the solution of the cubic equation gives $\alpha = 0.2142$ as the condition for lift-off from the walls, and the results of equations (4.7) are valid only for α greater than this value.

It is of interest that there is no value of α in the range $0.2142 < \alpha < 0.5$ for which the values of M_S and M_C become equal. Instead, the value of M_C decreases as α increases, eventually changing sign, while the value of M_S reaches a minimum for $\alpha = 0.3500$, for which value

$$[M_S]_{min} = (0.0873)\left(\frac{W\ell}{8}\right). \qquad (4.8)$$

According, therefore, to this elastic analysis, equation (4.8) results from the optimum arrangement of the brace supports if the bending moments in the beam are to be as small as possible. Figure 4.8(a) displays the results of equations (4.1) and (4.7), the values of the bending moments being plotted against the value of α. At $\alpha = 0.2071$ the values of M_S and M_C are equal, as has been noted, equation (4.3), and the ends of the beam are clear of the supporting walls. This 'undergraduate' optimum design does not depend so critically on the doubtful assumptions that have to be made for an elastic analysis, since the system is statically determinate. If the elastic assumptions are accepted, the ends of the beam touch down on the walls for $\alpha = 0.2142$, at which point a discontinuity may be seen in the slope of the curves in fig. 4.8(a), and the 'best' elastic design, equation (4.8), occurs for $\alpha = 0.3500$. Some typical bending-moment diagrams are sketched in fig. 4.9.

As a matter of interest, although not of present concern, it may be noted that in the range $0.3574 < \alpha < 0.4264$ the value of M_S does not correspond to the largest bending moment in the beam. Figure 4.10 gives a sketch of the bending-moment diagram for the end span, and the value of $[M]_{max}$ is

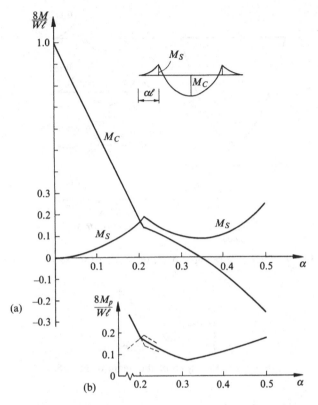

Fig. 4.8. (a) Elastic bending moments for fig. 4.6. (b) Plastic collapse analysis.

given by

$$[M]_{\max} = \frac{W\ell}{8}\left[1 - \frac{2M_S}{\alpha W\ell}\right]^2,$$ (4.9)

where the value of M_S is the known function of α, equation (4.7).

(c) A 'plastic' analysis

Wood is one of those ductile structural materials which (unlike glass or cast iron) is able to force a given construction into a 'load-sharing' mode. Moreover, since a plastic solution to the present problem is independent of the essentially unknowable conditions of support of the beam, it has at least the relevance and applicability of the elastic solution given above.

Figure 4.11 shows the three possible modes of collapse of the beam,

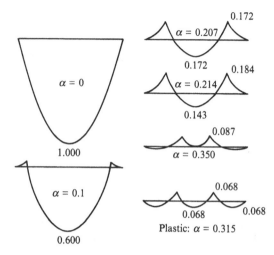

Fig. 4.9. Bending-moment diagrams for various values of α.

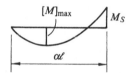

Fig. 4.10. End span of the beam of fig. 4.6.

depending on the value of α. For α small ($\alpha < 0.2071$) mode (a) will occur; this is the statically determinate lift-off case, for which $M_C > M_S$. For $\alpha > 0.2071$ the mode switches to that shown in fig. 4.11(b) and finally, as α is increased further, mode (c) occurs. The maximum design moments M_p from this plastic analysis are

$$
\left.
\begin{aligned}
\text{(a)} \qquad 0 < \alpha < 0.2071, \quad M_p &= (1 - 4\alpha)\frac{W\ell}{8}, \\[2mm]
\text{(b)} \quad 0.2071 < \alpha < 0.3153, \quad M_p &= \tfrac{1}{2}(1 - 2\alpha)^2\,\frac{W\ell}{8}, \\[2mm]
\text{(c)} \qquad 0.3153 < \alpha < 0.5, \quad M_p &= \left(0.6863\alpha^2\right)\frac{W\ell}{8}.
\end{aligned}
\right\} \qquad (4.10)
$$

The value $\alpha = 0.3153$ results from equating the second and third expressions for M_p.

Equations (4.10) for the 'plastic' analysis of the beam are plotted in

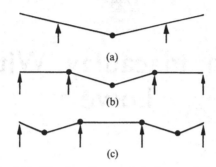

Fig. 4.11. Plastic collapse of the beam of fig. 4.6.

fig. 4.8(b); cf. fig. 4.8(a). For $\alpha < 0.2071$ the elastic and plastic analyses are of course identical for the statically determinate beam; for $\alpha > 0.2071$ the plastic analysis gives a value of bending moment M_p always less than the largest elastic moment for the same value of α. The optimum value of α for the plastic analysis is 0.3153, where the switch occurs between modes (b) and (c), and the corresponding value of the bending moment is

$$[M_p]_{\min} = (0.0682)\,\frac{W\ell}{8} \qquad (4.11)$$

(cf. equation (4.8)); the bending-moment diagram is sketched in fig. 4.9.

Both the (suspect) elastic analysis and the 'plastic' analysis indicate that some sort of best design would be achieved with braces provided to the beam at about one-third span ($\alpha = 0.35$ elastic, $\alpha = 0.32$ plastic). However, such braces would be long and might be uneconomical. At Durham the braces are positioned to provide support at a value of α roughly equal to 0.21; it will be recalled that at $\alpha = 0.2071$ the central and support bending moments are equal (equation (4.3)), while at $\alpha = 0.2142$ the side walls start to provide support for the beams. Thus the braces at Durham should ensure that the wall plates are largely relieved of direct bearing, the roof loads being transmitted down the braces to the feet of the wall posts. Indeed it is evident for some of the roof beams at Durham that their ends are barely in contact with the walls.

At St Albans the raking braces are positioned so that the value of α approaches 1/4.

Clebsch, Macaulay, Wittrick, Lowe

The function

$$w = \frac{F}{EI}\frac{z^3}{6},$$ (5.1)

where $z = (x - a)$, has the following properties:

$$
\left.
\begin{aligned}
&w = 0 \text{ for } x = a, \\[2mm]
&\frac{dw}{dx}\left(= \frac{F}{EI}\frac{z^2}{2}\right) = 0 \text{ for } x = a, \\[2mm]
&\frac{d^2w}{dx^2}\left(= \frac{F}{EI}z\right) = 0 \text{ for } x = a, \\[2mm]
\text{and}\quad &\frac{d^3w}{dx^3} = \frac{F}{EI}.
\end{aligned}
\right\}
$$ (5.2)

Similarly, the function

$$w = \frac{F}{P\alpha}(\alpha z - \sin \alpha z),$$ (5.3)

where $\alpha^2 = P/EI$ and $z = (x - a)$ as before, has the properties:

$$
\left.
\begin{aligned}
&w = 0 \text{ for } x = a, \\[2mm]
&\frac{dw}{dx}\left(= \frac{F}{P}(1 - \cos \alpha z)\right) = 0 \text{ for } x = a, \\[2mm]
&\frac{d^2w}{dx^2}\left(= \frac{\alpha F}{P}\sin \alpha z\right) = 0 \text{ for } x = a, \\[2mm]
\text{and}\quad &\frac{d^3w}{dx^3}\left(= \frac{F}{EI}\cos \alpha z\right) = \frac{F}{EI} \text{ for } x = a.
\end{aligned}
\right\}
$$ (5.4)

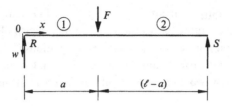

Fig. 5.1. Beam subject to transverse load.

Thus both of the functions of equations (5.1) and (5.3) are zero at $x = a$, and their first and second derivatives are also zero at that point. However, the third derivative in both cases has value F/EI.

The functions are of use in the solution of linear differential equations for which discontinuities can occur in the third derivatives. Equation (5.1) arises in the solution of the problem of an elastic beam having uniform flexural rigidity EI and subjected to a concentrated transverse load F. In fig. 5.1, the coordinate axes are taken at the left-hand end of the beam, and x measures the distance along the beam and w the downward deflexion. The point load F divides the beam into two regions, for which the differential equations governing the deflected form are

$$EI\frac{d^2 w_1}{dx^2} = -Rx, \qquad (5.5)$$

$$\text{and} \quad EI\frac{d^2 w_2}{dx^2} = -Rx + F(x - a). \qquad (5.6)$$

These two second-order differential equations will each have two constants of integration, and certain boundary conditions will obtain at the two ends of the beam which will give some information as to the values of these constants.

(Thus, in fig. 5.1, $w_1 = 0$ for $x = 0$ and $w_2 = 0$ for $x = \ell$. The beam has been sketched as simply supported, and the values of R and S may be found from the equations of statics, but the argument is in fact more general. Had the beam been clamped at its ends, for example, then the value of R would not be known from statics, and an initially unknown bending moment acting at the left-hand end of the beam would also enter the calculations. However, additional boundary conditions $dw_1/dx = 0$ at $x = 0$ and $dw_2/dx = 0$ at $x = \ell$ will eventually give sufficient information to determine the values of

R and of the clamping moment. It is shown later, in equations (5.25)–(5.27), that such a complex approach to the solution may be simplified.)

An essential requirement for the solution of equations (5.5) and (5.6) is that there should be geometrical continuity at $x = a$, that is, that the deflexions and slopes from the two equations should be identical at that point:

$$\left. \begin{array}{ll} \text{for } x = a, & w_1 = w_2 \\[2mm] \text{and} & \dfrac{dw_1}{dx} = \dfrac{dw_2}{dx}. \end{array} \right\} \tag{5.7}$$

Conditions (5.7) will give two further relations between the values of the four constants of integration.

Clebsch pointed out in 1862 (§87; the same paragraph numbering is adopted in Saint-Venant's French edition of 1883) that the continuity conditions (5.7) could be satisfied with the *same* complementary function $(\alpha x + \beta)$ for *both* equations (5.5) and (5.6) if they were integrated in the form

$$EIw_1 = -\frac{Rx^3}{6} + \alpha x + \beta, \tag{5.8}$$

$$EIw_2 = -\frac{Rx^3}{6} + \frac{F}{6}(x-a)^3 + \alpha x + \beta. \tag{5.9}$$

It is evident that equations (5.8) and (5.9) give the required continuity at $x = a$. By extension, the deflexions of a beam loaded by a number of concentrated loads may be written

$$\left. \begin{array}{l} EIw_1 = -\dfrac{Rx^3}{6} \hspace{5cm} +\alpha x + \beta, \\[4mm] EIw_2 = -\dfrac{Rx^3}{6} + \dfrac{F_1}{6}(x-a_1)^3 \hspace{2.5cm} +\alpha x + \beta, \\[4mm] \hspace{1cm} \vdots \\[4mm] EIw_n = -\dfrac{Rx^3}{6} + \dfrac{F_1}{6}(x-a_1)^3 + \dfrac{F_2}{6}(x-a_2)^3 + \cdots \\[4mm] \hspace{2cm} \cdots + \dfrac{F_{n-1}}{6}(x-a_{n-1})^3 \;\; +\alpha x + \beta. \end{array} \right\} \tag{5.10}$$

Macaulay introduced a 'curly bracket' notation in 1919 for this same problem of a transversely loaded beam:

$$\left.\begin{array}{rcl} \{f(x)\}_a & = & 0 \text{ for } x < a, \\ \{f(x)\}_a & = & f(x) \text{ for } x \geq a. \end{array}\right\} \tag{5.11}$$

(The precise specification of $\{f(x)\}_a$ at $x = a$ is important for the case when a suddenly applied couple acts on the beam.) With Macaulay's notation, equations (5.10) are replaced by the single equation

$$\begin{aligned} EIw_n = -\frac{Rx^3}{6} &+ \frac{F_1}{6}\{x - a_1\}_{a_1}^3 + \frac{F_2}{6}\{x - a_2\}_{a_2}^3 + \cdots \\ &\cdots + \frac{F_{n-1}}{6}\{x - a_{n-1}\}_{a_{n-1}}^3 + \alpha x + \beta, \end{aligned} \tag{5.12}$$

it being understood that, when any term within the brackets $\{\}$ becomes negative, the whole term is omitted.

Thus equation (5.12) implies the evaluation of only two constants of integration by reference to the external boundary conditions. Had separate bending equations been written for each of the n segments of the beam, then $2n$ constants of integration would have been involved. Macaulay's method was taught at Cambridge and popularized by Case (1925), and was of obvious help in the manual solution of problems in the bending of beams.

Wittrick took up the matter in 1965, and extended the analysis to deal with the problem of bending in the presence of axial load. (He also studied by the same technique the question of beams on elastic foundations, and certain problems in shells.) Finally Lowe made a thorough study of the problem in 1971, using the notation $z = (x - a)$; thus equation (5.12) may be written

$$w = -\frac{Rx^3}{6EI} + \frac{F_1}{6EI}z_1^3 + \frac{F_2}{6EI}z_2^3 + \cdots + \frac{F_{n-1}}{6EI}z_{n-1}^3 + \alpha x + \beta. \tag{5.13}$$

As before, if any variable z becomes negative, then the corresponding term is omitted from equation (5.13).

Equation (5.5) is integrable simply in quadratures, and the use of the variables z (or of Macaulay's brackets) seems self-evident. The matter is not so straightforward when bending with axial load is considered. Figure 5.2 shows the same beam loaded transversely, but now also with an end compressive load P; the bending equations for the two segments are

$$\left.\begin{array}{rcl} EI\dfrac{d^2w_1}{dx^2} + Pw_1 & = & -Rx, \\[4mm] \text{and}\quad EI\dfrac{d^2w_2}{dx^2} + Pw_2 & = & -Rx + F(x - a). \end{array}\right\} \tag{5.14}$$

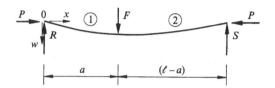

Fig. 5.2. The beam of fig. 5.1 with, additionally, end compressive load.

The question to be answered is how equations (5.14) may be solved most easily in order that the required degree of continuity is obtained at $x = a$, and, by extension, at $x = a_1, a_2, \ldots, a_{n-1}$ if several transverse loads are applied.

For the case of a uniform beam for which EI is constant, both of equations (5.14), and indeed all equations for segments of the beam between loads, can be reduced to the form

$$EI\frac{d^4w}{dx^4} + P\frac{d^2w}{dx^2} = 0, \tag{5.15}$$

or, setting $P/EI = \alpha^2$,

$$\frac{d^4w}{dx^4} + \alpha^2\frac{d^2w}{dx^2} = 0. \tag{5.16}$$

The complementary function of equation (5.16) is

$$w = A\cos\alpha x + B\sin\alpha x + Cx + D. \tag{5.17}$$

The particular integral of equation (5.16) appears to be $w = 0$, but Wittrick made use of the fact that a particular integral plus an arbitrary complementary function is also a particular solution of a linear differential equation of the type considered here. Thus, instead of $w = 0$, the particular integral of equation (5.16) will be taken to be

$$w = b\cos\alpha x + c\sin\alpha x + dx + e. \tag{5.18}$$

Now the *same* constants A, B, C and D of equation (5.17) will be used for *every* segment of the beam, exactly as the same constants α and β were used for the same problem in the absence of axial load. Thus the complementary function will be the same throughout the beam, but the constants b, c, d and e will be different for each segment. These four constants must be chosen to give the required conditions of continuity at $x = a$, namely that deflexion, slope and bending moment $(EI\,d^2w/dx^2)$ must be continuous, but that the shear force $(EI\,d^3w/dx^3)$ must jump by F. Thus, from equation (5.18),

$$(w)_a = b\cos\alpha a \qquad +c\sin\alpha a \qquad +da + e \;=0,$$

$$\left(\frac{dw}{dx}\right)_a = -\alpha b\sin\alpha a \qquad +\alpha c\cos\alpha a \quad +d \qquad = 0,$$

$$\left(\frac{d^2w}{dx^2}\right)_a = -\alpha^2 b\cos\alpha a \quad -\alpha^2 c\sin\alpha a \qquad\qquad = 0,$$

$$\left(\frac{d^3w}{dx^3}\right)_a = \alpha^3 b\sin\alpha a \qquad -\alpha^3 c\cos\alpha a \qquad\qquad = \frac{F}{EI},$$

(5.19)

from which

$$b = \frac{F}{\alpha^3 EI}\sin\alpha a, \quad c = -\frac{F}{\alpha^3 EI}\cos\alpha a,$$

$$d = \frac{F}{\alpha^2 EI}, \quad e = -\frac{Fa}{\alpha^2 EI}.$$

(5.20)

Hence equation (5.18) becomes

$$w = \frac{F}{\alpha^3 EI}\left[\alpha(x-a) - \sin\alpha(x-a)\right],\qquad(5.21)$$

$$\text{or } w = \frac{F}{\alpha^3 EI}(\alpha z - \sin\alpha z).\qquad(5.22)$$

The function of equation (5.22) is of course the function of equation (5.3).

Thus the complete general solution of a beam-column loaded transversely by a number of loads $F_1, F_2, \ldots$ may be written

$$\begin{aligned} w \;=\; & A\cos\alpha x + B\sin\alpha x + Cx + D \\ & +\frac{F_1}{\alpha^3 EI}(\alpha z_1 - \sin\alpha z_1) + \frac{F_2}{\alpha^3 EI}(\alpha z_2 - \sin\alpha z_2) + \cdots \end{aligned}\qquad(5.23)$$

where the four constants A, B, C and D are to be determined from the boundary conditions of the problem. For a simple support at the origin, for example, for which $w = 0$ and $d^2w/dx^2 = 0$, it will be seen that $A = D = 0$. For the problem of fig. 5.2, involving a single load F, the complete solution is

$$w = \frac{F}{\alpha^3 EI}\left[\frac{\sin\alpha(\ell - a)}{\sin\alpha\ell}\sin\alpha x - \frac{\alpha(\ell - a)}{\ell}x + (\alpha z - \sin\alpha z)\right].\qquad(5.24)$$

It will have been noted that by using the fourth-order differential equation (5.16) the need for preliminary statical analysis of the overall structure is obviated. For the solution of the beam problem in the absence of axial load, equation (5.13) assumes that the value of the reaction R is known. A more general approach is to start from the equation

$$\frac{d^4w}{dx^4} = p, \tag{5.25}$$

where p represents the intensity of any distributed transverse loading on the beam. The most general complementary function for equation (5.25) is

$$w = Ax^3 + Bx^2 + Cx + D, \tag{5.26}$$

so that equation (5.13) may be written in the alternative form

$$w = Ax^3 + Bx^2 + Cx + D + \frac{F_1}{6EI}z_1^3 + \frac{F_2}{6EI}z_2^3 + \cdots, \tag{5.27}$$

where, as in equation (5.23), the four arbitrary constants are to be found from the boundary conditions. For a clamped end at the origin, for example, $C = D = 0$, while for a pinned end, $B = D = 0$.

The elastica

The history of mathematics in the seventeenth century has been well studied, and indeed perhaps too much attention has been paid to the quarrel between Newton and Leibnitz, and their supporters, about the invention of the calculus. What is incontrovertible is the explosive influence that invention had on the development of applied mathematics and mechanics. By the middle of the next century, however, progress had been so rapid that only professional mathematicians can now contribute to the history of the subject. Newton's fluxions will be understood by the historian, but Euler's calculus of variations moved the subject to a different level of learning. There is a corresponding gap in the history of the development of mathematics in the eighteenth century, with some notable exceptions, such as those provided by Truesdell.

Euler's equation is, in fact, simple and easily memorable. If I is a functional of $y = y(x)$, where

$$I = \int_a^b F\left(x, y, y', y'', \ldots\right) dx,
\tag{6.1}$$

and the minimum of I is sought, then the required minimizing function y satisfies

$$\frac{\partial F}{\partial y} - \frac{d}{dx}\left(\frac{\partial F}{\partial y'}\right) + \frac{d^2}{dx^2}\left(\frac{\partial F}{\partial y''}\right) - \cdots = 0,
\tag{6.2}$$

where the right-hand side is zero on condition that sufficient end-point conditions are specified. (If equation (6.2) has two terms only, then $y(a)$ and $y(b)$ must be specified; if the equation has three terms, then $y'(a)$ and $y'(b)$ must also be specified, and so on.)

Daniel Bernoulli, one of perhaps a dozen mathematicians in the world who knew and appreciated what Euler had done, was quite capable of applying equation (6.2), in 1742, to the problem of the elastica. The difficulty was that Daniel Bernoulli, having obtained his fourth-order differential equation, could not solve it. He wrote to Euler on 20 October 1742 suggesting that

Euler himself should tackle the problem. Daniel Bernoulli had found that the 'vis viva potentialis', $\int ds/R^2$, where R is the radius of curvature of a bent elastic strip, was a minimum for the elastic curve of his uncle James. (The integral, the 'potential live force', is, to within a constant, the strain energy in bending.) He proposed to Euler that he should apply the calculus of variations to the inverse problem of finding the shape of the curve of given length connecting two given points, the slopes being specified at those points, so that $\int ds/R^2$ is a minimum.

Euler's response was to add an appendix (*Additamentum I, De Curvis Elasticis*) to a book on the calculus of variations (*Methodus inveniendi lineas curvas...*) published in 1744. He obtains the fourth-order differential equation without difficulty, and then in turn manages to integrate it until it is of first order, discusses exhaustively, without further explicit solution, all possible forms that the solution could take, and finally makes the last integration in the form of infinite series (representing the elliptic integrals).

If $p = y'$ and $q = p' = y''$, then the integral to be minimized is

$$I = \int \frac{ds}{R^2} = \int \frac{q^2}{(1+p^2)^{5/2}} dx \equiv \int Z \, dx, \text{ say,} \tag{6.3}$$

where the curvature has been introduced by the expression

$$\frac{1}{R} = \frac{q}{(1+p^2)^{3/2}}, \tag{6.4}$$

and where, of course, $ds = (1+p^2)^{1/2} dx$. Indeed, the length of the elastica is fixed, so that

$$J = \int ds = \int (1+p^2)^{1/2} dx \equiv \int F \, dx, \text{ say,} \tag{6.5}$$

is a constant. The expressions (6.3) and (6.5) involve the first and second differentials of y, so that application of equation (6.2) will result in an equation of three terms, requiring the specification of position and slope at each end of the elastica. The integrals in expressions (6.3) and (6.5) are evaluated between these boundaries.

A new variable $K = (I - \alpha J)$ is taken, where α is an arbitrary (Lagrangean) multiplier. It is known, since J is constant, that dJ is zero; if then dK is made to be zero, so also will be dI, and the required minimization will be achieved. Using equation (6.2),

$$-\frac{d}{dx}\left(\frac{\partial Z}{\partial p}\right) + \frac{d^2}{dx^2}\left(\frac{\partial Z}{\partial q}\right) + \alpha \frac{d}{dx}\left(\frac{\partial F}{\partial p}\right) = 0. \tag{6.6}$$

Now

$$\frac{\partial Z}{\partial p} = -\frac{5pq^2}{\left(1+p^2\right)^{7/2}} = P, \text{ say,}$$

$$\frac{\partial Z}{\partial q} = \frac{2q}{\left(1+p^2\right)^{5/2}} = Q, \text{ say,} \qquad (6.7)$$

and $\quad \dfrac{\partial F}{\partial p} = \dfrac{p}{\left(1+p^2\right)^{1/2}},$

so that

$$\frac{dP}{dx} - \frac{d^2Q}{dx^2} - \alpha\frac{d}{dx}\left[\frac{p}{\left(1+p^2\right)^{1/2}}\right] = 0, \qquad (6.8)$$

or

$$P - \frac{dQ}{dx} - \alpha\frac{p}{\left(1+p^2\right)^{1/2}} = \beta, \qquad (6.9)$$

where β is a constant of integration.

Equation (6.9) may be multiplied through by $dp = qdx$ to give

$$Pdp - qdQ - \alpha\frac{pdp}{\left(1+p^2\right)^{1/2}} = \beta dp, \qquad (6.10)$$

and, noting that $dZ = Pdp + Qdq$,

$$dZ - (Qdq + qdQ) = \alpha\frac{pdp}{\left(1+p^2\right)^{1/2}} + \beta dp, \qquad (6.11)$$

so that

$$\alpha\left(1+p^2\right)^{1/2} + \beta p + \gamma = Z - Qq = -\frac{q^2}{\left(1+p^2\right)^{5/2}}, \qquad (6.12)$$

where γ is another constant of integration.

Changing (arbitrarily) the signs of α, β and γ,

$$q = \frac{dp}{dx} = \left(1+p^2\right)^{5/4}\left[\alpha\left(1+p^2\right)^{1/2} + \beta p + \gamma\right]^{1/2}. \qquad (6.13)$$

Thus Euler has already performed two integrations on the initial fourth-order equation (6.6). Euler now notes(!) that

$$\frac{d}{dp}\left[\frac{2\left\{\alpha\left(1+p^2\right)^{1/2}+\beta p+\gamma\right\}^{1/2}}{\left(1+p^2\right)^{1/4}}\right]$$

$$=\frac{\beta-\gamma p}{\left(1+p^2\right)^{5/4}\left[\alpha\left(1+p^2\right)^{1/2}+\beta p+\gamma\right]^{1/2}}, \qquad (6.14)$$

so that, using equation (6.13),

$$\frac{d}{dp}[\;] = (\beta-\gamma p)\frac{dx}{dp}, \qquad (6.15)$$

that is,

$$\frac{2\left\{\alpha\left(1+p^2\right)^{1/2}+\beta p+\gamma\right\}^{1/2}}{\left(1+p^2\right)^{1/4}} = \beta x - \gamma y + \delta. \qquad (6.16)$$

The coordinate axes may be shifted and rotated without loss of generality so that, in effect, $\gamma = \delta = 0$, and equation (6.16) may then be solved for p in the form

$$p = \frac{dy}{dx} = \frac{n^2 x^2 - m a^2}{\left[n^2 a^4 - \left(n^2 x^2 - m a^2\right)^2\right]^{1/2}}, \qquad (6.17)$$

where $4m = \alpha a^2$ and $4n = \beta a^2$. The variable x occurs as x^2 in equation (6.17), and Euler moves the coordinates back to a general position to give

$$p = \frac{dy}{dx} = \frac{\alpha + \beta x + \gamma x^2}{\left[a^4 - \left(\alpha + \beta x + \gamma x^2\right)^2\right]^{1/2}}, \qquad (6.18)$$

where α, β and γ are now different constants from those originally assigned. Similarly, the arc lengths may be obtained from the equation

$$\frac{ds}{dx} = \frac{a^2}{\left[a^4 - \left(\alpha + \beta x + \gamma x^2\right)^2\right]^{1/2}}. \qquad (6.19)$$

Equations (6.18) and (6.19), derived by Euler in the way shown above on the basis of the variational calculus, are those which govern the form of the elastica, and are essentially those obtained by James Bernoulli in 1695. Euler interrupts his analysis at this point to show that equations (6.18) and (6.19) can be derived at once from Bernoulli's bending equation $M = EI/R$. In

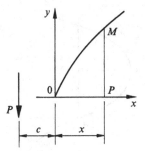

Fig. 6.1. The elastica bent by a single force P.

fig. 6.1 the elastica is bent by a single force P, so that, for a general point M,

$$P(c+x) = \frac{EI}{R} = -EI \frac{\dfrac{d^2y}{dx^2}}{\left[1 + \left(\dfrac{dy}{dx}\right)^2\right]^{3/2}}, \tag{6.20}$$

or

$$P\left(\frac{1}{2}x^2 + cx + f\right) = -EI \frac{\dfrac{dy}{dx}}{\left[1 + \left(\dfrac{dy}{dx}\right)^2\right]^{1/2}}, \tag{6.21}$$

from which

$$p = \frac{dy}{dx} = \frac{P\left(\frac{1}{2}x^2 + cx + f\right)}{\left[(EI)^2 - P^2\left(\frac{1}{2}x^2 + cx + f\right)^2\right]^{1/2}}, \tag{6.22}$$

and this is clearly of the same form as equation (6.18).

Euler then proceeds to classify, without numerical solution at this stage, the nine types of elastic curve, that is, he sketches graphically the elliptic integrals resulting from the integration of equation (6.18). For this purpose, Euler takes the form of equation (6.17),

$$\frac{dy}{dx} = \frac{\alpha + x^2}{\left[a^4 - (\alpha + x^2)^2\right]^{1/2}}, \tag{6.23}$$

in which the origin is at A in fig. 6.2, and the y axis is vertically down in the line of the applied force P. Comparison of the forms of equations (6.22) and

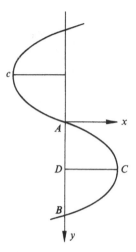

Fig. 6.2. General shape of the elastica (after Euler 1744).

(6.23) shows that

$$a^2 = \frac{2EI}{P}.$$ (6.24)

Euler sets $(a^2 - \alpha) = c^2$, so that equation (6.23) becomes

$$\frac{dy}{dx} = \frac{a^2 - c^2 + x^2}{\left[(c^2 - x^2)(2a^2 - c^2 + x^2)\right]^{1/2}},$$ (6.25)

and, further,

$$\frac{ds}{dx} = \frac{a^2}{\left[(c^2 - x^2)(2a^2 - c^2 + x^2)\right]^{1/2}}.$$ (6.26)

The sine of the angle of slope of the curve, dy/ds, therefore has value $(1 - c^2/a^2)$ at the origin A, while the tangent of the angle has value

$$\left[\frac{dy}{dx}\right]_{x=0} = \frac{a^2 - c^2}{c(2a^2 - c^2)^{1/2}}.$$ (6.27)

It will be seen from the denominator of equation (6.25) that x cannot exceed, numerically, the value c, so that the curve (say fig. 6.2) lies between $x = \pm c$. Moreover, at $x = \pm c$ the slope of the curve is infinite, so that the portion AC in fig. 6.2 may be sketched and, since the curve is an odd function of x, so may the portion Ac. Euler then deduces that the portion CB of the curve is

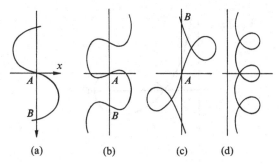

Fig. 6.3. Elastic curves of the second, fourth, sixth and eighth classes
(after Euler 1744).

a reflexion about the line CD of the portion CA, so that the whole, periodic,
curve may be constructed.

The precise form of the curve is clearly dependent on the ratio c/a. As
a final preliminary, Euler evaluates the length AC (that is, he performs the
quadrature of equation (6.26)) and the length AD (from equation (6.25)) in
terms of a series of powers of $(c/a)^2$. Thus

$$AC = \frac{\pi a}{2\sqrt{2}}\left\{ 1 + \left(\frac{1}{2}\right)^2\left(\frac{c^2}{2a^2}\right) + \left(\frac{1}{2}\right)^2\left(\frac{3}{4}\right)^2\left(\frac{c^2}{2a^2}\right)^2 \right.$$

$$\left. + \left(\frac{1}{2}\right)^2\left(\frac{3}{4}\right)^2\left(\frac{5}{6}\right)^2\left(\frac{c^2}{2a^2}\right)^3 + \cdots \right\}.$$

(6.28)

Euler distinguishes nine classes of solution, of which classes 2,4,6 and 8 are
sketched in fig. 6.3. Classes 3,5 and 7 are the transition curves between those
shown in fig. 6.3, and classes 1 and 9 represent the end points of the analysis.
Class 1 occurs for c very small, that is, for $c \ll a$. Since, as has been seen,
the maximum excursion of the curve is c, then x is also very small, and
equation (6.25) may be written

$$\frac{dy}{dx} = \frac{a}{\sqrt{2}\left(c^2 - x^2\right)^{1/2}}.$$

(6.29)

This has solution

$$x = c\sin\left(\frac{y\sqrt{2}}{a}\right),$$

(6.30)

so that, for very small values of c, the deflected form is sinusoidal. Further,

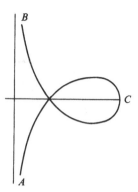

Fig. 6.4. Elastic curve of the seventh class (after Euler 1744).

from equation (6.28), the length AC is $\pi a/2\sqrt{2}$, and this is equal to $\ell/2$, where ℓ is the half-period of the sine wave. Thus, using equation (6.24),

$$\ell^2 = \frac{\pi a^2}{2} = \frac{\pi^2 EI}{P}, \tag{6.31}$$

and the 'Euler buckling load' $P_e = \pi^2 EI/\ell^2$ has been derived.

Class 2 has the general form sketched in fig. 6.3(a), and occurs for $c < a$; the load P necessary to maintain equilibrium is greater than P_e.

Class 3 ocurs for $c = a$; equation (6.27) shows that the slope of the curve is zero at the origin, and the load P is normal to the curve at that point. This is the rectangular elastica.

Class 4 is sketched in fig. 6.3(b), and ocurs for $1 < c^2/a^2 < 1.651868$. At the upper limit points A and B coincide, and the elastica takes on the form of a figure eight; this is Euler's Class 5.

In Class 6, fig. 6.3(c), the points A and B have crossed over, and the solution is valid for $c^2 < 2a^2$. It will be seen from equation (6.27) that if $c^2 > 2a^2$, the curve does not cut the y axis. For $c^2 = 2a^2$, which gives Euler's Class 7, the governing equation (6.25) becomes

$$\frac{dy}{dx} = \frac{x^2 - \tfrac{1}{2}c^2}{x\left(c^2 - x^2\right)^{1/2}}, \tag{6.32}$$

and this may be integrated to give

$$y = \left(c^2 - x^2\right)^{1/2} - \frac{1}{2}c\log\left\{\frac{c + \left(c^2 - x^2\right)^{1/2}}{x}\right\}. \tag{6.33}$$

This curve is sketched in fig. 6.4; the y axis is an asymptote.

For Class 8, $c^2 > 2a^2$, Euler sets $c^2 = 2a^2 + g^2$, so that equation (6.25) becomes

$$\frac{dy}{dx} = \frac{x^2 - \frac{1}{2}c^2 - \frac{1}{2}g^2}{[(c^2 - x^2)(x^2 - g^2)]^{1/2}}. \tag{6.34}$$

Thus the curve lies between $x = g$ and $x = c$, and its general form is that of fig. 6.3(d).

Class 9 is, effectively, that of pure bending, for which the elastica takes on the form of a circle; the case occurs for c infinite.

Euler's complete classification outlined above, without detailed solution of the first-order differential equation, is of extraordinary ingenuity. Class 1, giving the Euler buckling load of a pin-ended strut, is of course the result that has been of basic engineering importance. Euler returned to a discussion of this class in a later paper (1757) in which he derived directly the approximate 'engineering' equation

$$EI\frac{d^2y}{dx^2} = -Py. \tag{6.35}$$

Mechanisms of collapse

A 'frame' resists the action of external loads primarily by bending of its members. Thus the loads V and H applied to the plane rectangular portal frame of fig. 7.1(a) will give rise to certain bending moments. It is the evaluation of this single variable, the bending moment M, that is the objective of the structural analysis of the frame. Further, for the example of fig. 7.1, in which the loads are applied at discrete points and the members are straight between nodes, a knowledge of the values of the bending moment at those nodes (five in number in fig. 7.1) will give a complete description of the state of the frame. In fig. 7.1(b) the general bending-moment diagram has been sketched (with the sign convention that bending moments producing tension on the outside of the frame are positive).

The application of the first of the master equations of the theory of structures, that of statical equilibrium, will give some relations between the values of the bending moments at the nodes, M_A to M_E in fig. 7.1(b). As usual, a statical analysis may be made straightforwardly by the use of virtual work, and fig. 7.1(c) shows a pattern of deformation involving discontinuities at the five cardinal points. Between these hinges the members of the frame remain straight; the two columns are each inclined at a small angle θ to their original directions, and the two half-beams at an angle ϕ. The resulting hinge discontinuities marked in fig. 7.1(c) follow a similar sign convention to that for the bending moments.

It is of course evident that the mechanism of fig. 7.1(c) has more than one degree of freedom (i.e. motions described by θ and ϕ), and so cannot (except for a unique set of loads carried by the frame) correspond to any possible equilibrium state. It is, however, a state of possible deformation, and application of the equation of virtual work leads at once to

$$H(h\theta) + V\left(\tfrac{\ell}{2}\phi\right) = M_A(\theta) + M_B(\phi - \theta) + M_C(-2\phi)$$
$$+ M_D(\phi + \theta) + M_E(-\theta),$$

60

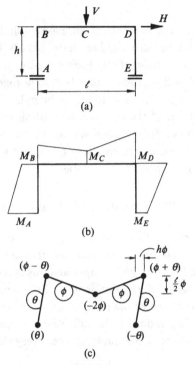

Fig. 7.1. Rectangular portal frame. (b) General bending-moment
diagram. (c) Most general deformation mechanism.

or $\quad (M_A - M_B + M_D - M_E - Hh)\theta$

$$+ \left(M_B - 2M_C + M_D - \frac{V\ell}{2} \right) \phi = 0. \qquad (7.1)$$

Now equation (7.1) must hold for any arbitrary values of θ and ϕ, so that

$$\left. \begin{array}{llll} M_A - & M_B & +M_D & -M_E & = Hh \\ & M_B & -2M_C & +M_D & = V\dfrac{\ell}{2}. \end{array} \right\} \qquad (7.2)$$

and

Two independent equilibrium equations have therefore been established, and there are no other patterns of deformation (similar to that of fig. 7.1(c)) which will lead to any further equation. Indeed, the portal frame of fig. 7.1 is an example of the well-known form of the fixed-base arch, which has three degrees of statical indeterminacy; no other independent equations can be found from statics.

To make further progress with the analysis use must be made of the other two master equations. For example, if an elastic solution is required, then material properties will be introduced to relate, linearly, the bending moment and curvature at every section of the frame; finally, geometrical considerations ('compatibility') will require the members to fit together (for example at the corners B and D in fig. 7.1), and to satisfy boundary conditions (clamped ends at the feet A and E of the frame). The analysis is straightforward, but the equations may be complex; moreover, as was noted in chapter 1, very small variations in the boundary conditions can lead to very large changes in the elastic bending moments in the frame.

7.1 Plastic analysis of plane frames

By contrast, simple plastic theory cuts away many of these difficulties, at least for the plane frame (as will be seen, matters are less straightforward for the space frame). If the plastic collapse analysis is required for the frame of fig. 7.1, then the basic information is given by the equilibrium equations (7.2). Material properties enter the analysis by the statement that there is a largest value of bending moment, the full plastic moment M_0, that can be sustained by the members of the frame, so that everywhere

$$|M_i| \leq (M_0)_i \tag{7.3}$$

(M_0 may of course be allowed to take different values at different sections of the frame). Finally, the geometrical statement is simply that there should be a mechanism of collapse; such a mechanism will be formed quite independently of any imperfections of connexions between members, or conditions at the foundations.

As is well known, a mechanism of plastic collapse corresponds to a 'breakdown' of an equilibrium equation, such as the first of equations (7.2). Indeed, if M_A is set equal to M_0 (assumed to be constant for this simple example), M_B set equal to $-M_0$, and so on, so that the left-hand side of the equation is maximized, then the collapse equation $Hh = 4M_0$ is obtained, with the corresponding mechanism of fig. 7.2(a). (This is the 'θ' mechanism of fig. 7.1(c), with ϕ set equal to zero.) Similarly, the second of equations (7.2) gives $V\ell/2 = 4M_0$, corresponding to the mechanism of fig. 7.2(b). (As a matter of interest, there is a third mechanism of collapse − and no other mechanism is possible for positive V and H − given by the superimposition of equations (7.2). This leads to the equation $Hh + V\ell/2 = 6M_0$ and the mechanism of fig. 7.2(c). It will be clear that this last mechanism

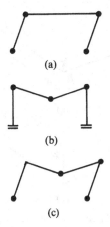

(a)

(b)

(c)

Fig. 7.2. The three possible mechanisms of collapse for the fixed-base
portal frame.

is not independent of the other two, and this fact has led to the powerful
'combination of mechanisms' method for plastic analysis, either by hand or
by computer.)

A count of the variables involved gives a more general view of the plastic
collapse of plane frames under *proportional loading*. For proportional loading,
all loads acting on the frame (V and H in fig. 7.1) are specified in terms
of one of their number, W, say; the value of W is required at collapse (or,
what is the same thing, the value of the collapse load factor is required). The
frame has N critical sections at which hinges might form (five in number in
fig. 7.1), so that there are

$$(N + 1) \text{ unknown quantities } W, M_i, \quad i = A, \ldots, N.$$

The information available for the solution comes from

$$(N - R) \quad \text{equilibrium equations connecting } W, M_i$$
$$\text{and } (H) \quad \text{yield conditions } |M_i| = M_0,$$

where H is the number of hinges involved in the collapse mechanism and R
is the number of redundancies. If the information available is to solve the
problem, then

$$(N - R) + (H) = (N + 1),$$

$$\text{or} \quad H = (R + 1). \tag{7.4}$$

The notion of a *regular* collapse mechanism has been derived, in which the

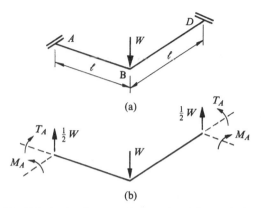

Fig. 7.3. A simple space frame: the right-angle bent.

number of hinges exceeds by one the number of redundancies in the frame. Thus the portal frame of fig. 7.1(a) has three redundancies, and a regular collapse mechanism for the frame has four hinges; see figs. 7.2(a) and (c).

A partial collapse mechanism may form in a frame, for which the number H of hinges is fewer than $(R+1)$. The mechanism of fig. 7.2(b) involves only collapse of the beam at a calculable value of the applied load V, and one redundancy (in the columns) remains for the frame as a whole if it collapses in this partial mode. Indeed, a partial mode is the most likely collapse mechanism for any practical frame, and the plastic designer may well have some difficulty in completing a statical analysis so that, for example, stability of columns may be checked. (The elastic designer is not worried by this problem, since the results from a computer analysis will give a tidy set of bending moments throughout the frame. However, the problem is still there, as will have been noted from the remarks in chapter 1.)

7.2 A simple space frame

A special type of space frame will be discussed in which all members of the frame lie in a plane, and all loads act perpendicular to that plane. A grillage of beams forms a frame of this type. In establishing equilibrium equations for such structures, moments may be taken about axes lying in the plane, but no information results from taking moments about an axis parallel to the loads.

The simplest possible type of such a space frame is sketched in fig. 7.3,

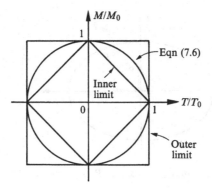

Fig. 7.4. Yield surface for a plastic hinge formed under bending and twisting.

in which two equal members AB and BD, continuous at B, are built in to fixed walls at A and D so that the frame lies in a horizontal plane. This right-angle bent carries a single vertical load W at B. The loading will induce both bending moments M and twisting moments T in the members, and it is to be expected that, following the ideas of simple plastic theory for plane frames, collapse will occur by the formation of hinges at the ends A and D.

If the value of full plastic moment in the absence of twisting moment is M_0, and of the full plastic twisting moment in the absence of bending is T_0, then it will be assumed that a hinge will form in general when

$$f\left(\frac{M}{M_0}, \frac{T}{T_0}\right) = \text{const.} \tag{7.5}$$

Equation (7.4) represents a *yield surface* in plasticity theory, and its form is severely limited by the convexity requirement of that theory. If, for example, the yield surface is doubly symmetric, which is very often the case in practice, then equation (7.5) must lie between the limits sketched in fig. 7.4. Also shown in fig. 7.4 is the circle

$$\left(\frac{M}{M_0}\right)^2 + \left(\frac{T}{T_0}\right)^2 = 1, \tag{7.6}$$

and many practical members, for example tubes of circular or rectangular section, will have yield criteria approximating to equation (7.6). For the purpose of the present discussion, equation (7.6) will be taken to govern the formation of plastic hinges in the frame.

At collapse of the bent, fig. 7.3(b), the hinge at A will be acted on by a

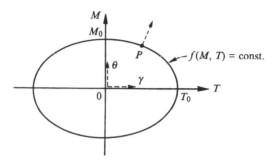

Fig. 7.5. The deformation vector at P is normal to the yield surface.

bending moment M_A and twisting moment T_A, and (M_A, T_A) will satisfy

$$\left(\frac{M_A}{M_0}\right)^2 + \left(\frac{T_A}{T_0}\right)^2 = 1. \tag{7.7}$$

By symmetry, the same values of bending and twisting moments will occur at the hinge at D, as marked in fig. 7.3(b). A single equation of statics may be written which relates the quantities marked in the figure:

$$M_A + T_A = \tfrac{1}{2}W\ell. \tag{7.8}$$

Equations (7.7) and (7.8) are the only two equations which can be written using information derived from equilibrium and the plastic properties of the material, and they do not between them provide enough information to evaluate the collapse load W.

There is thus a profound difference between the plastic analysis of this simple space frame and corresponding analyses of plane frames. Figure 7.3 is indeed a collapse mechanism (and it is in fact the correct mechanism), but the analysis is not statically determinate. The third equation of the theory of structures must be used: the geometry of deformation of the collapse mechanism must be examined.

The extra necessary information is provided by the *normality condition* of plasticity theory. In fig. 7.5 a general yield surface is sketched for the formation of a plastic hinge under combined bending and twisting. At point P on this yield surface the variables M and T satisfy the yield condition

$$f(M, T) = \text{const.} \tag{7.9}$$

Superimposed on the axes of M and T in fig. 7.5 are axes for θ and γ, measuring respectively the angle of plastic bending and the angle of plastic

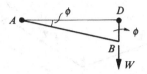

Fig. 7.6. Collapse mechanism for the right-angle bent of fig. 7.3.

twist at the hinge. Then the *normality condition* states that the vector of deformation (θ, γ) shown at P must be normal to the yield surface at that point, that is

$$\frac{\theta}{\gamma} = \frac{\partial f / \partial M}{\partial f / \partial T}. \tag{7.10}$$

If, then, the circular condition of equation (7.6) applies, equation (7.10) requires that

$$\frac{\theta}{\gamma} = \left(\frac{T_0}{M_0}\right)^2 \left(\frac{M}{T}\right). \tag{7.11}$$

Figure 7.6 shows a side elevation of the bent in its collapsing state; a bending rotation ϕ at hinge A implies a twisting rotation of the same magnitude ϕ at hinge D. The bending rotation at D has, of course, the same value ϕ, so that, at both hinges, quation (7.11) gives

$$\frac{M_A}{T_A} = \frac{M_0^2}{T_0^2}. \tag{7.12}$$

Equations (7.7) and (7.12) may now be solved simultaneously to give

$$M_A = \frac{M_0^2}{\left(M_0^2 + T_0^2\right)^{\frac{1}{2}}}, \quad T_A = \frac{T_0^2}{\left(M_0^2 + T_0^2\right)^{\frac{1}{2}}}, \tag{7.13}$$

so that, finally, equation (7.8) gives an expression for the collapse load

$$\tfrac{1}{2} W \ell = \left(M_0^2 + T_0^2\right)^{\frac{1}{2}}. \tag{7.14}$$

This elaborate procedure is necessary even for this simple frame, and the same steps may be followed for more complex examples, as will be seen. However, it may be noted that, just as for the plane frame, the collapse load, equation (7.14), may be obtained by examining the breakdown of an equilibrium equation. Thus the single equilibrium equation (7.8) gives the general value of the load W in terms of the moment and torque acting at hinge A; if the value of the load W is maximized subject to the yield condition, equation (7.7), then precisely equation (7.14) is recovered.

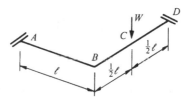

Fig. 7.7. The same bent under different loading.

7.3 An 'overcomplete' mechanism

The problem shown in fig. 7.7 is very much more difficult, despite its superficial resemblance to that of fig. 7.3. For ease of calculation the truly circular criterion

$$M^2 + T^2 = M_0^2 \tag{7.15}$$

will be used; that is, it will be assumed that $M_0 = T_0$. The flow rule associated with equation (7.15) is therefore

$$\frac{\theta}{\gamma} = \frac{M}{T}; \tag{7.16}$$

cf. equation (7.11).

The difficulty with the problem of fig. 7.7 lies in the determination of the correct collapse mechanism. As a first trial, the mechanism of fig. 7.3(b) will be assumed, for which $\theta = \gamma = \phi$ at the hinges; see fig. 7.6. From equations (7.16), $T = M$ at each hinge, and equation (7.15) then gives

$$T = M = \frac{1}{\sqrt{2}} M_0. \tag{7.17}$$

The work equation for the collapse mechanism of fig. 7.6 (with the load W acting at C, the midpoint of BD), gives

$$W \left(\frac{1}{2} \ell \phi \right) = 2 (T \phi + M \phi),$$

$$\text{that is} \quad W = 4\sqrt{2} \frac{M_0}{\ell}. \tag{7.18}$$

The usual theorems of plasticity theory apply — the value of W given by equation (7.18) is an upper bound on the true value at collapse. A statical examination of the frame shows at once that the yield condition, equation (7.15), is violated at the loading point C.

It would seem likely, therefore, that the true collapse mechanism will involve a hinge at C, but a simple beam mechanism, with hinges at B, C and

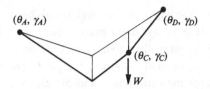

Fig. 7.8. Collapse mechanism for fig. 7.7.

D, is quickly seen to be incorrect. There is no other arrangement of hinges, that includes a hinge at C, which gives a mechanism of the usual kind of one degree of freedom. The actual mode of collapse is one which involves hinges at the three locations A, C and D; see fig. 7.8. At each of these hinge points two degrees of freedom are permitted; since the original frame had three redundancies, the resulting mechanism, viewed purely as a mechanism, appears to have three degrees of freedom. However, the flow rule must be obeyed at each hinge, and exactly the right number of equations is obtained for the solution of the problem. A count of the variables involved shows how this comes about.

7.4 A count on the space frame

The arguments which led up to equation (7.4) may be followed for the space frame. If the frame is collapsing with H hinges, then the numbers of unknown quantities in the analysis are

$(2H + 1)$ unknown forces $W, M_i, T_i,$

and $(2H - 1)$ unknown hinge rotations $\theta_i, \gamma_i,$

(since the collapse mechanism must be able to experience an arbitrary displacement). The information for the solution of the problem comes from

$(2H - R)$ equilibrium equations involving $W, M_i, T_i,$

(H) yield conditions $f(M_i, T_i) = $ const.,

(H) flow equations $\theta_i / \gamma_i = (\partial f / \partial M)_i / (\partial f / \partial T)_i,$

and (R) geometrical relationships connecting $\theta_i, \gamma_i.$

Thus there are precisely $4H$ equations which are available to determine $4H$ unknowns, but no rule has emerged which leads to the number of hinges required for collapse of a space frame.

Indeed, a 'regular' collapse mechanism, by analogy with the example of the plane frame, would be of one degree of freedom, for which all the hinge discontinuities (both bending and twisting) could be written in terms of a single parameter. If now one extra plastic hinge were inserted into this 'regular' collapse mechanism, then four extra unknowns (M, T, θ, γ) would be introduced into the problem. However, the yield condition and the flow rule may be written for the extra hinge, and by taking moments about two axes through the hinge, two extra equilibrium equations may also be written.

Unlikely as it may seem, therefore, the 'overcomplete' mechanism of fig. 7.8 is correct for the collapse of this particular right-angle bent. The collapse load, as a matter of interest, is given by

$$W = \frac{16}{\sqrt{10}} \frac{M_0}{\ell}, \qquad (7.19)$$

and the geometrical relationships between the rotations at the hinges have simple forms, such as $\theta_C = \theta_D - \gamma_A$; see fig. 7.8.

7.5 The work equation

Equation (7.18) resulted from a simple work balance for an assumed mechanism of collapse. It led to an upper bound on the value of the collapse load, that is, it gave an 'unsafe' estimate of the strength of the frame. Nevertheless, assumed patterns of hinge rotations often lead to reasonably accurate estimates of strength. If, for example, the value of θ_D in fig. 7.8 is set (arbitrarily) equal to unity, and (reasonable) values are assigned to the other five quantities in the figure, then a work equation may be written equating the work done by the load to the work dissipated in the hinges.

If the hinges form according to the circular criterion, equation (7.6), then the flow rule relates bending to twist at the hinge, equation (7.11). The two equations may be solved simultaneously to give the values of M and T, so that the work dissipated at a plastic hinge, $(M\theta + T\gamma)$, may be evaluated as

$$\left(M_0^2 \theta^2 + T_0^2 \gamma^2\right)^{\frac{1}{2}}. \qquad (7.20)$$

7.6 The plane frame: bending and axial load

The complexities in the analysis of space frames arise from the fact that plasticity develops under the action of more than one 'force' − two 'forces' were involved (bending moment and twisting moment) for the very restricted type of frame examined above. A more general space frame might be analysed on the basis that plastic hinges developed under bending about two

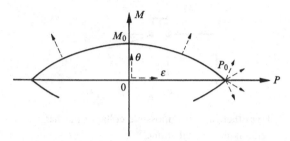

Fig. 7.9. Yield surface for a plastic hinge formed under bending and thrust.

axes as well as twisting; three variables would then be involved in the yield condition. The yield surface could be imagined in three-dimensional space, and the conditions of normality and expressions for the flow rule would still hold – as they would in a hyperspace of four dimensions, which would be needed if the effect of a single axial load had also to be considered at a hinge.

The designer of plane frames is accustomed to making allowance, if necessary, for the effect of axial load on the formation of 'one-dimensional' plastic hinges. If such a hinge forms under the combined action of a bending moment M and a thrust P, then the yield condition takes the form

$$f(M, P) \equiv \left(\frac{M}{M_0}\right) + \left(\frac{P}{P_0}\right)^n = 1, \qquad (7.21)$$

where P_0 is the value of the 'squash load' in the absence of bending. The exponent n in equation (7.21) has value 2 for a rectangular cross-section, and may approach unity for an I-section. In either case, the general form of the yield surface is as sketched in fig. 7.9.

In the 'engineering' analysis of frames, the values of full plastic moment at the hinge points are simply reduced from M_0 in accordance with equation (7.21); otherwise, the calculations proceed exactly as for the statically determinate case. However, the analysis appears to be no longer strictly 'one-dimensional'; the yield surface of equation (7.21) and fig. 7.9 implies that a rotation θ at a hinge must be accompanied by a stretch ε. It should be possible, therefore, to find collapse mechanisms of an unusual kind, in which hinges, formed under combined bending and axial load, give rise to both bending and stretching deformation.

Such a hypothetical (and, as will be seen, impossible) mechanism for the simple portal frame of fig. 7.1 is sketched in fig. 7.10. Each of the two hinges has a rotation θ; the stretch at hinge C is $\varepsilon_C = h\theta$, while the stretch at hinge

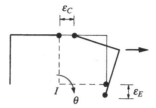

Fig. 7.10. Hypothetical but impossible collapse mechanism for the
rectangular portal frame.

E is $\varepsilon_E = -\frac{1}{2}\ell\theta$. Thus, from the normality condition indicated in fig. 7.9, a
tensile force must act at C and a compressive force at E. At either hinge the
numerical ratio ε/θ is of the order of the overall dimensions of the frame.

Such ratios are not possible with the yield condition of equation (7.21);
the curve sketched in fig. 7.9 is very shallow, and the ratio ε/θ given by the
normal vector is much less than a typical dimension of the frame. This is not
a matter of a misleading scale in the plot of fig. 7.9; it arises directly from
the flow rule derived from equation (7.21):

$$\frac{\varepsilon}{\theta} = \left(\frac{\partial f/\partial P}{\partial f/\partial M}\right) = n\left(\frac{P}{P_0}\right)^{n-1}\left(\frac{M_0}{P_0}\right). \tag{7.22}$$

The maximum value of ε/θ from equation (7.22) is clearly of order M_0/P_0
for $1 < n < 2$, and M_0/P_0 is of order d, where d is the depth of the structural
member. Indeed, for a symmetrical cross-section, the maximum value of ε/θ
from equation (7.22) is precisely $\frac{1}{2}d$ as P approaches the value P_0; the zero-
stress axis is always confined to lie within the depth of the member (except
for the special case $P = P_0$, where the usual fan of deformation occurs at
the vertex of the yield surface, fig. 7.9).

Mechanisms such as that sketched in fig. 7.10 can occur for plane frames
if the centres of rotation of the hinges (I in fig. 7.10) lie well *outside* the
depths of the numbers, whereas the flow condition requires that the centres
of rotation lie *inside* the depths of the members. While the conclusion is
general, it is sharpened for the usual case in which the depths of structural
members of plane frames are considered to be infinitesimal compared with
their overall dimensions, so that centre-line geometry is used in the analysis.
In this case only infinitesimal axial movement is possible; reduced values
of plastic moments can be calculated in the usual 'engineering' way from
equation (7.21), and the flow rule gives no information either to help the
analysis or to render it invalid.

The absolute minimum-weight design of frames

The word 'frame' (or 'beam') implies, as usual, that the primary structural action is that of bending. Secondary actions, such as those of shear and of axial load, may have to be taken into account in design – certainly their effects must be checked. However, the first task of a designer is to establish a distribution of bending moments for the structure. Thereafter, calculations may proceed in one of two basic ways. If a plastic design is being made, then each member will be designed so that its full plastic moment just exceeds, at each critical section, the value of the bending moment established by the designer for that section. If an elastic approach is used, then the stresses at each section must be kept within certain specified limiting values.

The initial bending-moment distribution established by the designer is, of course, not unique; it is one of infinitely many that can be found for a hyperstatic structure. However, as has been seen in chapter 1, both the elastic and the plastic approach outlined above are, by the plastic theorems, safe. The plastic design is, in general, direct, whereas the elastic design is iterative, since the bending-moment distribution is affected by the elastic properties of the members, and these are not known *a priori*.

Prismatic members are commonly used in steelwork design, since it is inconvenient, except for heavy fabricated construction, to vary the cross-section of a member within its length. The use of reinforced concrete gives more possibility of variation of cross-section, but in both steel and reinforced-concrete design some material will be 'wasted', since the full potential strength of a member (whether assessed elastically or plastically) will be realized at only a few cross-sections. It is clear that, at least in theory, if a cross-section could be varied continuously along the length of a member so that its strength exactly matched that required by the equilibrium bending-moment distribution, then some sort of minimum-weight design would be achieved. The minimum would in fact only be relative, since such a design would have been matched to a particular, and perhaps arbitrary, set of equilibrium moments. However, among this class of designs there will be one, and possibly more than one, equilibrium distribution which gives an absolute minimum weight.

This 'absolute minimum' design may not be easy to fabricate, and is likely to be more expensive than a design which wastes material; however, the study is of interest for its own sake, and because it leads to a target against which a practical design can be judged.

Thus if the bending moment at any section x of a beam or frame is denoted M, designs will be examined for which the strength of the member at that section is $|M|$. The 'strength' will not be defined for the time being — that is, it could be measured in terms of full plastic moment or in terms of a limiting elastic stress. It will be assumed that the total weight X of the frame is a linear function of the strength, so that, to some scale,

$$X = \int |M| dx, \tag{8.1}$$

where the integration extends over the length of all members. The bending-moment distribution which gives the absolute minimum weight will be denoted $|M^*|$, and

$$X^* = \int |M^*| dx. \tag{8.2}$$

The beam in fig. 8.1(a) carries a set of arbitrary (downward) loads, and the ends are supposed to be restrained so that clamping moments can act there. Had the beam been simply supported, then the bending moments M_w would be statically determinate (fig. 8.1(b)). As it is, the bending moments M_w are the free bending moments for the beam, and the actual bending moments of fig. 8.1(c) are determined by superimposing a reactant line. Thus, from fig. 8.1(c),

$$M = M_w - (A + Bx), \tag{8.3}$$

where the two constants A and B determine the position of the reactant line. Variation of the values of these two constants leads to different designs $|M|$ of the beam, and the minimum-weight design $|M^*|$ is sought.

It will be seen that, as sketched, two inflexion points occur (defined by the

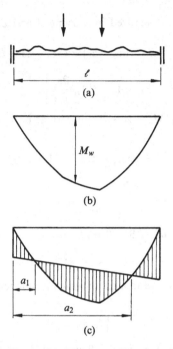

Fig. 8.1. Fixed-ended beam under general loading.

parameters a_1 and a_2) at which the bending moment changes sign. Since the strength of the beam follows exactly the distribution of bending moments M, this implies that the section of the beam will vanish at two locations; in practice, some material must be provided to transmit shear forces.

From equations (8.1) and (8.3), the weight of the design is

$$
\begin{aligned}
X &= \int_0^\ell |M_w - A - Bx| \, dx \\
&= \int_0^{a_1} (A + Bx - M_w) \, dx + \int_{a_1}^{a_2} (M_w - A - Bx) \, dx \\
&\quad + \int_{a_2}^\ell (A + Bx - M_w) \, dx \\
&= \int_0^\ell (A + Bx - M_w) \, dx + 2\int_{a_1}^{a_2} (M_w - A - Bx) \, dx. \quad (8.4)
\end{aligned}
$$

The value of X is to be minimized by varying A and B. Thus

$$\frac{\partial X}{\partial A} = \ell + 2 \left[-(a_2 - a_1) - (M_w - A - Bx)_{a_1} \frac{\partial a_1}{\partial A} \right.$$

$$\left. + (M_w - A - Bx)_{a_2} \frac{\partial a_2}{\partial A} \right]. \tag{8.5}$$

Now the value of $M = (M_w - A - Bx)$ is zero at both $x = a_1$ and $x = a_2$, so that

$$(M_w - A - Bx)_{a_1} = (M_w - A - Bx)_{a_2} = 0, \tag{8.6}$$

and the condition $\partial X/\partial A = 0$ gives

$$a_2 - a_1 = \tfrac{1}{2}\ell. \tag{8.7}$$

Similarly, the condition $\partial X/\partial B = 0$ leads to

$$a_2^2 - a_1^2 = \tfrac{1}{2}\ell^2, \tag{8.8}$$

and these two equations lead to a result which is independent of the arbitrary applied loading:

$$a_1 = \tfrac{1}{4}\ell, \; a_2 = \tfrac{3}{4}\ell. \tag{8.9}$$

Thus the beam loaded as in fig. 8.2(a) will have the minimum-weight bending-moment distribution sketched in fig. 8.2(b). The beam loaded as in fig. 8.2(c) will have the minimum-weight design of fig. 8.2(d), in which the beam has degenerated into a cantilever.

The results implied by equation (8.9) are remarkable. The minimum-weight design of any single-span beam is that sketched in fig. 8.2(a), and consists of a simply supported span of length $\tfrac{1}{2}\ell$ suspended from two end cantilevers each of length $\tfrac{1}{4}\ell$. The actual proportions of the cross-section of the beam are determined by the loading in a particular case, and of course by the chosen design procedure (e.g. limiting elastic stress or plastic collapse). However, the basic result is independent of the material properties, and arises from the linear assumption of equation (8.1).

The result was obtained by mechanical differentiation of equation (8.4), which related to a specific problem, but the matter may be viewed with more structural insight by the use of the equation of virtual work. The design $|M|$ of the frame, leading to the total weight of equation (8.1), is based upon considerations of equilibrium. As usual, the bending moments M may be related to the external loading by consideration of a compatible pattern of deformation. Thus, if changes of curvature κ are imposed on the frame, and

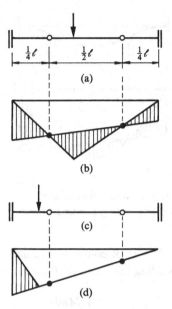

Fig. 8.2. Minimum-weight configuration for the fixed-ended beam.

the external loading is introduced by means of a set of free bending moments M_w (cf. fig. 8.1(b)), then

$$\int M_w \kappa dx = \int M \kappa dx = I \text{ (say)}. \tag{8.10}$$

Curvatures κ will be said to be *conformable* with bending moments M if sgn M = sgn κ everywhere.

The following theorem holds: If, for given loads, a design $|M^*|$ can be found that is conformable with a deflected form of the frame having constant curvatures $\kappa = \pm\kappa_0$, with no discontinuities of slope at inflexion points, then the design $|M^*|$ has minimum weight.

If indeed, a conformable deformation $\kappa = \pm\kappa_0$ is imposed on the putative minimum-weight design M^*, then equation (8.10) gives

$$I_0 = \int M^* \kappa dx = \int |M^*|\kappa_0 dx = \kappa_0 \int |M^*| dx. \tag{8.11}$$

If the weight equation (8.2) is now introduced, then

$$X^* = \frac{I_0}{\kappa_0}. \tag{8.12}$$

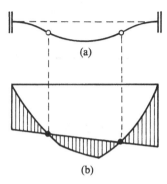

Fig. 8.3. (a) Deflexion configuration of the minimum-weight beam.
(b) The corresponding bending-moment diagram.

However, the quantity I_0 may of course be evaluated for any other equilibrium distribution of bending moments M, from equation (8.10), that is

$$I_0 = \int M\kappa dx, \tag{8.13}$$

where $\kappa = \pm\kappa_0$ as before.

Combining equations (8.12) and (8.13),

$$X^* = \int \frac{M\kappa}{\kappa_0}dx \le \int \frac{|M|\kappa_0}{\kappa_0}dx, \tag{8.14}$$

the inequality being necessary since the general distribution M will not in general be conformable with the imposed curvatures $\pm\kappa_0$. Equation (8.14) gives

$$X^* \le \int |M|dx = X, \tag{8.15}$$

that is, the weight X^* cannot exceed the weight X of any other design, and is therefore the minimum weight. (The requirement of no discontinuity of slope at inflexion points is necessary to prevent extra terms arising in equation (8.15). Since M and κ are in general not conformable, a hinging discontinuity θ_i would give rise to a virtual work term $M_i\theta_i$, and this could invalidate the inequality.)

The minimum-weight theorem, as stated above, requires the bending moments of the minimum-weight design to be conformable with a deflected form of the frame consisting of arcs of constant curvature. This geometrical requirement imposes a unique configuration on the fixed-ended beam — the

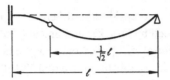

Fig. 8.4. Deflexions of the minimum-weight propped cantilever.

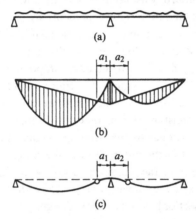

Fig. 8.5. Minimum-weight design of a two-span beam.

arcs sketched in fig. 8.3(a) are the only ones that will fit together without discontinuity of slope at the inflexion points. Since the signs of the bending moments must be the same as the signs of the curvatures everywhere in the beam, the reactant line is located immediately, fig. 8.3(b). The same uniqueness of solution is apparent for the propped cantilever, fig. 8.4; the position of the point of inflexion is independent of the loading system.

This independence holds only for the simplest beams and frames. The two-span beam on simple supports, for example (fig. 8.5(a)), will have a bending-moment diagram of the form shown in fig. 8.5(b), and a deflected conformable shape (with arcs of constant curvature) shown in fig. 8.5(c). The geometry of this deflected shape will provide one relationship between the distances a_1 and a_2 which define the positions of the inflexion points; a second relationship must be found from the particular bending-moment diagram. It is possible to establish certain general geometrical relationships to help with the solution of such problems, and numerical techniques of solution have been established.

The minimum-weight theorem discussed here is a special case of a more general theorem established by Drucker and Shield (1957) for the continuum. The general theorem, however, applies only to plastic structures, whereas the use of virtual work in establishing equation (8.15) made no immediate reference to material properties. Indeed, there is no necessity for the frame actually to deflect into a shape consisting of arcs of constant curvature. This displacement pattern was used in the equation of virtual work to establish regions for which the bending moments are 'hogging' and 'sagging', and to locate the inflexion points. The frame is then designed so that its strength at each section follows exactly the minimum-weight bending-moment distribution (and is zero at the inflexion points).

For an elastic design, the strength constraint would be the attainment of a maximum limiting stress. If the beam were of rectangular section, for example, and of constant depth, then the width at each cross-section should be made proportional to the bending moment there — the limiting stress would be reached at the outer fibres of the beam at all cross-sections. Further, the total weight of the beam would satisfy the linear assumption of equation (8.1). These same arguments apply also (with the introduction of a shape factor for the section) to fully plastic design.

Inverse design of grillages

A structure will in general be hyperstatic, so that its elastic analysis requires the simultaneous solution of all three of the master equations. A continuous beam, for example, may rest on several supports, and initially unknown redundant forces will act on the system. Such a beam may carry different loads in adjacent spans, and, as a consequence, the cross-section of the beam may well vary from span to span (while perhaps being prismatic within each span). There are no formal difficulties in the elastic calculations associated with this continuous beam; the loads, the section properties and the boundary conditions could be introduced into a computer program, for example, and the required results will be produced.

The *design* of such a beam, as opposed to its analysis, is not so straightforward; the section properties cannot be introduced numerically into the calculations, since it is precisely the determination of the section (or sections) of the beam which is the object of the design process. Design proceeds, in fact, by trial and error, whether this process is done manually or by computer. A guess is made of the cross-sectional properties of the members; an analysis is made to determine stresses and deflexions throughout the beam; and finally a check is made as to whether the criteria of strength and stiffness are satisfied.

In this trial-and-error process, one design variable can be allowed, namely a scale factor. If, for example, all the section properties were increased in proportion, then all the numerical results would be decreased in proportion. However, this single 'degree of freedom' will allow only one of the structural criteria to be satisfied; if the stress condition governs, then the structure may be too stiff; if deflexions are given their limiting values, then the structure may be understressed. Moreover, in either case limiting conditions will be reached at isolated portions of the structure, and it may be a complex problem to alter the sections of the design so that larger portions of the structure contribute to the satisfactory and economic performance of the whole.

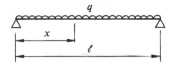

Fig. 9.1. Simply supported beam.

9.1 Inverse elastic design

An inverse method of elastic design can allow criteria of strength and stiffness to be satisfied simultaneously. More importantly, tapering members may be used without difficulty, since the mathematical analysis of members of continuously varying section is not needed. To continue the discussion with reference to beams, the steps in the inverse method are:

1. If w is the deflexion at a section x of the beam, then a deflected form $w = f(x)$ is assumed. The function w satisfies the boundary conditions of slope and deflexion.
2. The curvature $\kappa = d^2w/dx^2$ is then calculated for each section. The bending moment M at each section can now be written as $M = -EI\kappa$, where EI is an unknown flexural rigidity.
3. The bending moments M must satisfy the equilibrium equations; thus, since M and κ are known, the flexural rigidity EI may be determined.
4. The depth of the beam may be fixed at each section so that the stress does not exceed the permitted value.

It will be appreciated that the conventional process involves a double integration of the basic equation $\left(EI\,d^2w/dx^2 = -M\right)$; the inverse process involves a double differentiation $\left(\kappa = d^2w/dx^2\right)$. Two simple examples will illustrate the above steps.

9.2 A simply supported beam

Figure 9.1 shows a simply supported beam carrying a uniformly distributed load q per unit length. The inverse design is started by assuming a deflected form, say

$$w = \frac{4w_0}{\ell^2}x(\ell - x). \tag{9.1}$$

Deflexions are zero at both ends of the beam, and the maximum deflexion is w_0. From equation (9.1),

$$\kappa = \frac{d^2w}{dx^2} = -\frac{8w_0}{\ell^2}. \tag{9.2}$$

The curvature is thus constant for this assumed deflexion function. If a beam of constant depth h were used, then the upper and lower fibres of the beam would have the same stress at all cross-sections. For a symmetrical section, if the permitted maximum stress is σ_0, then the depth h is given by

$$\frac{\sigma_0}{\frac{1}{2}h} = E\frac{8w_0}{\ell^2}, \tag{9.3}$$

that is

$$h = \frac{\sigma_0}{E}\frac{\ell^2}{4w_0}. \tag{9.4}$$

The bending moment at any section is given by

$$M = -EI\kappa = EI\frac{8w_0}{\ell^2}. \tag{9.5}$$

However, the bending moment M must also satisfy the equilibrium equation

$$\frac{d^2M}{dx^2} + q = 0, \tag{9.6}$$

that is,

$$M = -\frac{q}{2}\left(x^2 + Ax + B\right), \tag{9.7}$$

or, making use of the conditions $M = 0$ for $x = 0, \ell$,

$$M = \frac{q}{2}x(\ell - x). \tag{9.8}$$

Equation (9.8) could of course have been written directly for this simple statically determinate problem; the differential equation (9.6) applies to *any* system, determinate or hyperstatic.

Equations (9.5) and (9.8) give

$$EI\frac{8w_0}{\ell^2} = \frac{q}{2}x(\ell - x), \tag{9.9}$$

and the flexural rigidity EI is evidently required to vary parabolically for this design. Equations (9.3) and (9.9) combine to give

$$I = \frac{qh}{4\sigma_o}x(\ell - x). \tag{9.10}$$

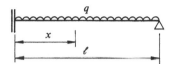

Fig. 9.2. Propped cantilever.

If, for example, the beam is made from concrete having a rectangular cross-section of variable width b and constant depth h, then equation (9.10) leads to

$$b = \frac{3}{4}\frac{q\ell^2}{\sigma_0 h^2}\frac{4x}{\ell}\left(1 - \frac{x}{\ell}\right) = b_0\frac{4x}{\ell}\left(1 - \frac{x}{\ell}\right). \qquad (9.11)$$

As a numerical example, a reinforced-concrete beam will be designed for $\ell = 15m$ to carry a load $q = 30$ kN/m. Nominal properties for the concrete are $\sigma_0 = 10$ N/mm² and $E = 25$ kN/mm². The maximum deflexion is limited to $w_0 = 20$ mm. The calculations will be inexact since the simple theory of bending has been used to derive the above equations, and no allowance has been made for the effects of reinforcement, shift of neutral axis, and so on.

Equation (9.4) gives first $h = 1125$ mm, and equation (9.11) then gives $b_0 = 400$ mm. The beam tapers parabolically from 400 mm width at the centre to zero at each end; no allowance has been made for the effect of shearing force.

9.3 A propped cantilever

The same beam is now to be designed as a propped cantilever, fig. 9.2. The assumed deflexion function must satisfy the condition of zero displacement at both ends of the beam, and zero slope at $x = 0$. A simple function with these properties is

$$w = \frac{8w_0}{\ell^3}x^2(\ell - x); \qquad (9.12)$$

the deflexion at midspan is w_0, and the maximum deflexion in the span is about 19 per cent greater. The curvature is given by

$$\kappa = \frac{d^2w}{dx^2} = \frac{16w_0}{\ell^3}(\ell - 3x). \qquad (9.13)$$

Equation (9.13) shows that the largest curvature in the span has value $32w_0/\ell^2$; if a beam of constant depth h is used, then the required value is

$$h = \frac{\sigma_o}{E}\frac{\ell^2}{16w_0}, \tag{9.14}$$

and the maximum stress σ_0 will be reached at only one cross-section of the beam.

If the same steps are followed as for the simply supported beam, then

$$M = -EI\frac{16w_0}{\ell^3}(\ell - 3x) \equiv -\frac{q}{2}(x^2 + Ax + B). \tag{9.15}$$

The constants A and B which result from the equilibrium equation must be chosen to give the required inflexion point at $x = \frac{1}{3}\ell$, and, since the bending moment must be zero at $x = \ell$, then

$$\left(x^2 + Ax + B\right) \equiv \frac{1}{3}(\ell - 3x)(\ell - x). \tag{9.16}$$

Finally, therefore, equation (9.15) gives

$$EI = \frac{q\ell^4}{96w_0}\left(1 - \frac{x}{\ell}\right). \tag{9.17}$$

The flexural rigidity of this beam (of constant depth) decreases uniformly from the left-hand end to zero at the right-hand end.

This design is likely to be uneconomical since, as noted, a beam of constant depth will be fully stressed at only one cross-section. If the constriction of constant depth is maintained, then the beams of constant curvature discussed in chapter 8 may be used − for the particular case of the propped cantilever, the deformation pattern of fig. 8.4 gives the required design. The bending-moment diagram must be arranged to give the indicated inflexion point.

However, this particular problem will not be discussed further, although it is straightforward to extend the ideas to multi-span beams, and to deal with complex patterns of loading − indeed, bending-moment distributions can be determined graphically rather than analytically. The examples of simple beams have been given as an introduction to the problem of the inverse design of grillages − it will be seen that analogous steps may be applied to such two-dimensional systems of beams.

9.4 Elastic design of grillages

The inverse design of grillages starts, as for the problem of beams, with the assumption of a pattern of deflexion. The analysis then proceeds by using small-deflexion theory for flat plates. When the flexural rigidity of a flat plate has been determined, to correspond with a given loading, the plate is replaced by beams forming a grillage of a specific shape. The material of the plate is 'concentrated' along the lines of the beams, each beam replacing some width of the notional plate. Thus, if a certain flexural rigidity *per unit width* of the plate, say EI_0, has been calculated, and the beams are spaced at a distance ℓ in that region of the plate, then each beam must have flexural rigidity $EI_0\ell$ to carry the given loading. With a large number of beams in the grillage, little error will be introduced by this replacement of the theoretically continuous plate by discrete beams; even if the number of beams is small, the calculations will be reasonably accurate. The approximation is analagous to the replacement of a distributed load by a number of concentrated loads.

The deflexion of the plate will be taken to be $w = w(x, y)$ with respect to fixed axes in the plane of the plate. Then curvatures at any point are given by

$$\kappa_x = \frac{\partial^2 w}{\partial x^2}; \; \kappa_y = \frac{d^2 w}{\partial y^2}; \; \kappa_{xy} = \frac{\partial^2 w}{\partial x \partial y}. \tag{9.18}$$

The curvature κ_{xy} is a twisting curvature. If new axes (ξ, η) are taken which make an angle ψ with the axes (x, y), then similar expressions for curvature may be written:

$$\kappa_\xi = \frac{\partial^2 w}{\partial \xi^2}; \; \kappa_\eta = \frac{\partial^2 w}{\partial \eta^2}; \; \kappa_{\xi\eta} = \frac{\partial^2 w}{\partial \xi \partial \eta}. \tag{9.19}$$

These new expressions (9.19) are related to those referred to the original axes. In particular, the condition for the twisting curvature $\kappa_{\xi\eta}$ to be zero is

$$\cot 2\psi = \frac{\dfrac{\partial^2 w}{\partial x^2} - \dfrac{\partial^2 w}{\partial y^2}}{2\dfrac{\partial^2 w}{\partial x \partial y}} = f(x, y), \quad \text{say.} \tag{9.20}$$

If the substitution $\tan \psi = dy/dx$ is made, then equation (9.20) becomes

$$\frac{dy}{dx} = -f \pm \sqrt{1 + f^2}. \tag{9.21}$$

The two equations (9.21) give a set of orthogonal curves; along these curves the twisting curvature is zero, and they are in fact curves of principal curvature, whose values are

$$\frac{\partial^2 w}{\partial \xi^2}, \frac{\partial^2 w}{\partial \eta^2} = \frac{1}{2}\left(\frac{\partial^2 w}{\partial x^2} + \frac{\partial^2 w}{\partial y^2}\right) \pm \frac{\partial^2 w}{\partial x \partial y}\sqrt{1 + f^2}. \tag{9.22}$$

The inverse elastic design of grillages proceeds as follows.

1. A deflected form of the grillage $w(x, y)$ is assumed which satisfies the boundary conditions.
2. The function f is determined from equation (9.20), and equations (9.21) are solved to give the lines of principal curvature. The beams of the grillage will be run along these lines; these beams will not be subject to twisting, since it has been ensured that $\kappa_{\xi\eta} = 0$.
3. The principal curvatures are calculated from equations (9.22). From the values of these principal curvatures, the bending moments per unit width of the beams, M_ξ and M_η, can be determined:

$$M_\xi = -EI_\xi \frac{\partial^2 w}{\partial \xi^2}; \quad M_\eta = -EI_\eta \frac{\partial^2 w}{\partial \eta^2}. \tag{9.23}$$

   The flexural stiffness of the beam running in the ξ direction has been denoted EI_ξ; if a (variable) rectangular cross-section having depth h_ξ is used, for example, then $I_\xi = \frac{1}{12} b_\xi h_\xi^3$, where b_ξ is the width of the beam *per unit width of the plate*, measured perpendicular to the direction of ξ.
4. As for the one-dimensional beams, the depths h_ξ and h_η of the beams of the grillage may be assigned so that a given stress σ_0 is not exceeded.
5. Finally, the equilibrium equation must be satisfied. If q is the load per unit area carried by the grillage, then

$$\frac{\partial^2 M_\xi}{\partial \xi^2} + \frac{\partial^2 M_\eta}{\partial \eta^2} + q = 0. \tag{9.24}$$

This equation provides one relation between the two flexural stiffnesses EI_ξ and EI_η, on substitution of equation (9.23). The coordinates ξ and η are of course curvilinear, and equation (9.24) must be interpreted properly; in practice, it may be simpler to refer the equation to the original axes (x, y).

Some simple examples will make the process clear.

9.5 An infinite floor supported by columns

Figure 9.3(a) shows a floor supported by columns spaced regularly at a distance $2a$ apart in the x direction and $2b$ in the y direction. A typical panel is shown in fig. 9.3(b), with axes at the centre of the panel. The assumed deflexion function is

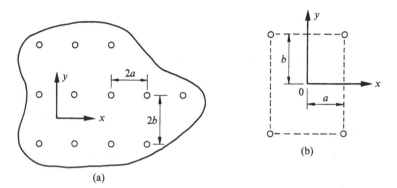

Fig. 9.3. Array of columns supporting a floor.

$$w = w_0 \left(1 - \frac{x^2}{a^2} - \frac{y^2}{b^2} + \frac{x^4}{2a^4} + \frac{y^4}{2b^4} \right). \tag{9.25}$$

The maximum deflexion w_0 occurs for $x = y = 0$; the deflexion at $x = \pm a$, $y = \pm b$ (that is, at the columns) is zero. Further, the slope $\partial w/\partial x$ is zero for $x = 0$ and $x = \pm a$, as it should be by symmetry, and similarly for the slope in the y direction. From equation (9.25),

$$\left.\begin{aligned} \frac{\partial^2 w}{\partial x^2} &= -\frac{2w_0}{a^2} \left(1 - 3\frac{x^2}{a^2} \right), \\[2mm] \frac{\partial^2 w}{\partial y^2} &= -\frac{2w_0}{b^2} \left(1 - 3\frac{y^2}{b^2} \right), \\[2mm] \frac{\partial^2 w}{\partial x \partial y} &= 0. \end{aligned}\right\} \tag{9.26}$$

The last of equations (9.26) shows that the lines of principal curvature are straight and parallel to the coordinate axes — a rectangular grillage of beams will be used.

From equations (9.23),

$$\left.\begin{aligned} M_x &= EI_x \frac{2w_0}{a^2} \left(1 - 3\frac{x^2}{a^2} \right) \\[2mm] \text{and} \quad M_y &= EI_y \frac{2w_0}{b^2} \left(1 - 3\frac{y^2}{b^2} \right), \end{aligned}\right\} \tag{9.27}$$

and these equations must satisfy the equilibrium equation (9.24), where, for this particular example, $(\xi, \eta) \equiv (x, y)$. If I_x and I_y are taken to be constant, then equation (9.24) gives

$$12w_0 E \left(\frac{I_x}{a^4} + \frac{I_y}{b^4} \right) = q, \qquad (9.28)$$

and any values of I_x and I_y satisfying this equation will give a satisfactory design.

The assumed deflexion function, equation (9.25), is valid for $-a \le x \le a$ and $-b \le y \le b$. Beams must be provided at the edges of this rectangle to transmit the load into the columns. For example, the beams of the grillage running in the x direction will be subject to a shearing force per unit width of value

$$Q_x = -\frac{\partial M_x}{\partial x} = \frac{12w_0}{a^4} EI_x.x, \qquad (9.29)$$

so that the edge beam at $x = a$ must carry a force per unit length of $2(12w_0 EI_x/a^3)$, where the factor of 2 arises from a similar shear force delivered by the adjacent panel. Thus the edge beam of length $2b$ carries a total uniformly distributed load of $4b \left(12w_0 EI_x/a^3\right)$, and it too may be designed by an inverse method.

9.6 A clamped circular grillage

A circular area of radius a is to carry a uniformly distributed load q per unit area; the beams of the grillage are fixed against both deflexion and rotation on the boundary. The deflexion function

$$w = \frac{w_0}{a^4} \left(a^2 - x^2 - y^2\right)^2 \qquad (9.30)$$

has the required boundary properties, and the curvatures are given by

$$\left. \begin{array}{l} \dfrac{\partial^2 w}{\partial x^2} = -\dfrac{4w_0}{a^4} \left(a^2 - 3x^2 - y^2\right), \\[2ex] \dfrac{\partial^2 w}{\partial y^2} = -\dfrac{4w_0}{a^4} \left(a^2 - x^2 - 3y^2\right), \\[2ex] \dfrac{\partial^2 w}{\partial x \partial y} = \dfrac{8w_0}{a^4} xy. \end{array} \right\} \qquad (9.31)$$

The function $f(x, y)$ defined by equation (9.20) is

$$f = \frac{x^2 - y^2}{2xy}, \qquad (9.32)$$

$2a$

Fig. 9.4. Possible design of a clamped circular grillage.

and equations (9.21) become

$$\frac{dy}{dx} = \frac{y}{x} \text{ or } -\frac{x}{y}. \tag{9.33}$$

Equations (9.33) integrate to give

$$\left.\begin{array}{r} y = kx, \\ x^2 + y^2 = c^2, \end{array}\right\} \tag{9.34}$$

as the lines of zero twisting; the orthogonal set of beams in the grillage are radii and concentric circles, as shown in fig. 9.4. Equations (9.22) give the principal curvatures along these lines:

$$\left.\begin{array}{rl} \dfrac{\partial^2 w}{\partial \xi^2} &= \dfrac{4w_0}{a^4}\left(3x^2 + 3y^2 - a^2\right), \\[2mm] \text{and} \quad \dfrac{\partial^2 w}{\partial \eta^2} &= -\dfrac{4w_0}{a^4}\left(a^2 - x^2 - y^2\right), \end{array}\right\} \tag{9.35}$$

from which, as usual, the depths of the beams may be calculated.

It is convenient to work in polar coordinates; the principal curvatures are

$$\left.\begin{array}{c} -\dfrac{4w_0}{a^4}\left(a^2 - 3r^2\right) \\[2mm] \text{and} \quad -\dfrac{4w_0}{a^4}\left(a^2 - r^2\right), \end{array}\right\} \tag{9.36}$$

from which

$$\left.\begin{array}{rl} M_r &= EI_r \dfrac{4w_0}{a^4}\left(a^2 - 3r^2\right) \\[2mm] \text{and} \quad M_\theta &= EI_\theta \dfrac{4w_0}{a^4}\left(a^2 - r^2\right). \end{array}\right\} \tag{9.37}$$

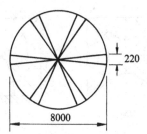

Fig. 9.5. Design of beams to cover a circular area.

Equation (9.24) may be written (taking account of symmetry):

$$\frac{d^2 M_r}{dr^2} + \frac{2}{r}\frac{dM_r}{dr} - \frac{1}{r}\frac{dM_\theta}{dr} = -q. \tag{9.38}$$

As a specific and very simple example, if $I_\theta = 0$, so that there are only radial beams in the design, then equations (9.37) and (9.38) combine to give

$$I_r = \frac{qa^4}{72Ew_0}. \tag{9.39}$$

If these beams are in reinforced concrete ($E = 25$ kN/mm^2 and $\sigma_0 = 10$ N/mm^2 as before), with $a = 4$ m, $w_0 = 5$ mm and $q = 5$ kN/m^2, then the first of equations (9.36) gives a maximum curvature $8w_0/a^2$ from which $h = 320$ mm if the permitted stress is just reached. Equation (9.39) then gives $I_r = 142 \times 10^3$ mm^3, from which $b_r = 0.0521$ mm/mm. At the periphery, the total width of the beams is therefore $(0.0521)(8000\pi) = 1309$ mm, say six beams each 220 mm wide. These beams taper uniformly to zero at the centre of the circular area; the arrangement is shown in fig. 9.5.

9.7 Square panels

Figure 9.6 gives sketch results for the elastic design of grillages for square panels. The function

$$w = \frac{w_0}{a^4}\left(a^2 - x^2\right)\left(a^2 - y^2\right) \tag{9.40}$$

is a suitable deflexion pattern for a grillage simply supported at the edges of a square panel of side-length $2a$. The equations giving the lines of zero twist must be integrated numerically, and the resulting shape of the grillage is sketched in fig. 9.6(a).

 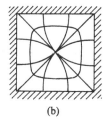

(a)   (b)

Fig. 9.6. Designs of grillages to cover a square area with ends of
beams (a) simply supported and (b) clamped.

Figure 9.6(b) shows a grillage, clamped at the edges of the panel, derived
from the function

$$w = \frac{w_0}{a^8} \left(a^2 - x^2\right)^2 \left(a^2 - y^2\right)^2. \tag{9.41}$$

9.8 Plastic design

The design process can be much easier if the designer is satisfied that strength
will be the governing criterion. The calculation of deflexions plays no part in
simple plastic design, and much of the work involved in the inverse method
described above can be avoided.

For grillages, for example, the beams may be run along orthogonal curves
of zero twist, as for the inverse elastic design. However, once equation (9.24)
is satisfied by any combination of the bending moments M_ξ and M_η, then the
beams may be designed to have strengths not less than these values; *any* solution of the equilibrium equation is satisfactory, and the bending moments are
no longer related to curvatures (as they are, for example, by equations (9.23)).

The problem is to determine the orthogonal curves of zero twist, and
this can certainly be done by assuming a deflexion function satisfying the
boundary conditions. (If this function involves inflexion points then, as usual,
the bending moments must change sign at these points.) For example, if a
grillage of straight beams is used, then

$$\frac{\partial^2 w}{\partial x \partial y} = 0, \tag{9.42}$$

and this equation has a general solution

$$w = g(x) + h(y) \tag{9.43}$$

where g and h are arbitrary functions.

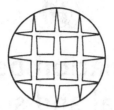

Fig. 9.7. Alternative grillage to cover a circular area.

Thus the deflexion function

$$w = \frac{w_0}{a^2} \left(a^2 - x^2 - y^2\right) \tag{9.44}$$

is of the form of equation (9.43), and has the value zero round a circle of radius a. The principal curvatures are constant, that is, no inflexion points occur within the circle; equation (9.44) is therefore suitable for the generation of a grillage of simply supported straight beams to cover a circular area. *One solution of the equilibrium equation (9.24) is*

$$M_x = M_y = \frac{q}{4} \left(a^2 - x^2 - y^2\right). \tag{9.45}$$

If these values of M_x and M_y are taken as the full plastic moments of the beams, then the resulting grillage, with three beams in each direction, is shown schematically in fig. 9.7.

It may be noted that, if an arbitrary set of orthogonal curves is sketched, then the value of $\cot 2\psi$ in equation (9.20) will be known at each point of the curvilinear grillage. Equation (9.20) will then in general define some deflexion function w having the sketched set of orthogonal curves as lines of principal curvature. It must be checked that this function w satisfies the boundary conditions of the problem but, with care, almost any 'reasonable' form of orthogonal grillage may be sketched by the designer (such as those of figs. 9.6(a), (b), which were derived elastically). This is a remarkable feature of the plastic approach; such orthogonal systems have been used in reinforced-concrete construction, notably by Nervi (1956).

The relation between incremental and static plastic collapse

A structure will in practice be acted upon by a number of independent loads (superimposed floor loads, snow, wind, crane loads, and so on). Moreover, each load will vary between limits which will be specified – the wind may blow from east to west, or not at all, or from west to east. The engineer making an elastic check of a given design will arrange for calculations to be made separately for each load; for any particular cross-section, the value of each load is then chosen to give the greatest and least action at that section. The designer is then able to determine the range of stress at the cross-section due to all the loads, and to make an assessment, according to given elastic rules of design, of the safety of the structure.

By contrast, the engineer making an estimate of the static plastic carrying capacity of a framed structure must arrange all the loading in the way expected to be most critical before the single plastic-collapse calculation is made. In practice there is no great difficulty in arriving at the worst combination of loads (although even for the simple portal frame the most critical position of a crane crab is not immediately obvious).

However, if loads on a frame do act randomly and independently within specified limits, then there is the possibility of incremental collapse of the frame. Moreover, it is shown below that incremental collapse is always a more critical phenomenon than the corresponding static collapse of the frame under the worst combination of steady loads having their maximum values.

Incremental collapse occurs in the following way. Under a certain combination of the independently varying loads it may be that a plastic hinge develops at one or more cross-sections of the frame. If so, then some small amount of rotation will take place at the hinge positions; the amount will be small, because a single hinge, or two or three hinges, will not in general be sufficient to form an overall or local mechanism of collapse. Any plastic deformation that does occur will be constrained by those portions of the frame that remain elastic. Under a different combination of loads it is possible that another hinge or hinges could form at different cross-sections. A yet further combination of loads could lead to yet further hinges, and so on.

Now, although collapse will not occur under any of these individual combinations of loads, it may be that had all the hinges formed simultaneously, then they would have given rise to a mechanism of collapse. If this is so, then incremental collapse is possible. A first combination of loads will lead to small irrecoverable hinge rotations; when the combination is removed, the frame is left slightly deformed. A second combination will lead to permanent deformations elsewhere in the frame, and so on. If the loading combinations follow in a more or less cyclic order, then, after a very few such cycles, the frame will exhibit pronounced overall deformation and will look, in fact, as if it were failing by a plastic collapse mechanism of the usual static type.

This sort of incremental collapse will occur if the magnitudes of the applied loads exceed certain values (to be calculated). For smaller values of the loads, then it is possible that some plastic deformation could occur in the first few cycles of loading, but that after a time all further variation of load would be resisted purely elastically by the frame. If this happens, then the frame is said to have shaken down under its specified set of variable loads.

The limit factor (that is, the shakedown limit), which divides incremental collapse from shakedown behaviour, can be calculated. In order to assign a numerical value to the limit, it is convenient to retain the idea of a load factor applied to the specified values of the loads. This load factor is applied not to the current value of a load, but to the range within which that load acts. Thus the limits $W^{\min}$ and $W^{\max}$ might represent the least and greatest value of a particular load W specified in an appropriate Code, so that

$$W^{\min} \leq W \leq W^{\max}. \tag{10.1}$$

The range of loading is now widened, hypothetically, by the factor λ, so that

$$\lambda W^{\min} \leq W \leq \lambda W^{\max}. \tag{10.2}$$

The largest value, λ_{S}, of the load factor at which the frame is unable to shake down is sought.

It is clear that for small enough values of λ there will be no plastic deformation as the loads vary within their individual ranges. As the value of λ is imagined to be slowly increased, then a state will be reached at which a single critical section (or, by accident of geometry, two or more critical sections) will just begin to yield under the most unfavourable combination of the independent loads. The corresponding value of λ is the conventional safety factor of stress calculated by the elastic designer. However, just as for the static case, the attainment of the yield stress at one or more cross-sections does not imply collapse; the value of λ may be increased and yet the frame might still be able to shake down. If the value of λ is above that of the conventional safety factor then yield is, by definition, occurring somewhere in the frame. Such yield implies in turn that, if all the loads were removed, the frame would no longer be stress free; the frame has been distorted, and residual (self-stressing) moments will have been induced. It is the presence of these residual moments m that allows the frame to shake down for values of λ below λ_S but above the value of the conventional safety factor.

The restriction on the value of λ may be calculated as follows. An elastic analysis must first be made, and the elastic bending moment $\mathcal{M}_i$ determined for each critical section i of the frame; this elastic analysis is made in the usual way on the assumption that the frame is initially stress free. The values of $\mathcal{M}_i$ will be functions of the loads; as individual loads vary between their prescribed limits, inequalities (10.1), so the bending moment $\mathcal{M}_i$ will vary between greatest and least values, say $\mathcal{M}_i^{\max}$ and $\mathcal{M}_i^{\min}$. With the value of load factor applied to the limits on the loads, inequalities (10.2), then the corresponding values of the bending moments are $\lambda\mathcal{M}_i^{\max}$ and $\lambda\mathcal{M}_i^{\min}$. These factored values of the elastic bending moments are those that would occur for the stress-free frame; in general, there will be a residual moment m_i at the cross-section which has been induced by plastic deformation in the first few cycles of loading, and which must be added to the elastic moments to give the total actual moment at the cross-section. Then the necessary conditions for shakedown to occur are that

$$\left.\begin{array}{l} \lambda\mathcal{M}_i^{\max} + m_i \le \left(M_p\right)_i, \\ \lambda\mathcal{M}_i^{\min} + m_i \ge -\left(M_p\right)_i. \end{array}\right\} \tag{10.3}$$

The counterpart to the usual lower-bound theorem for static loading is the statement that if *any* set of residual moments m_i can be found to satisfy inequalities (10.3) for every cross-section of the frame at a given value λ of the load factor, then shakedown will occur at that load factor. That is, the inequalities are both sufficient and necessary for the occurrence of shakedown.

Indeed, inequalities such as (10.3) can be written for every cross-section of the frame, and it is then a purely formal, if tedious, exercise to determine the largest value of λ, namely λ_S, for which they can be satisfied. Such an analysis would automatically indicate which of the inequalities were 'critical' at the value λ_S, and would thus indicate the incremental collapse mechanism. However, just as for static collapse, an 'upper-bound' approach through the examination of mechanisms will lead to simpler calculations.

Suppose that the correct mechanism for incremental collapse is known; that is, there is a pattern of hinges which, if they formed simultaneously, would give a mechanism of the usual type. The *virtual* mechanism will now be examined which has the same hinge locations as the incremental collapse mechanism, but for which all deformation occurs only at the hinges, the members of the frame itself remaining straight. A simple mechanism of this sort is sketched in fig. 2.5(b); the general mechanism under discussion has hinge rotations θ_i. If the value of θ_i at a particular section is positive, say θ_i^+, then at incipient incremental collapse of the frame the hinge at that section will form when the total bending moment reaches the value $+\left(M_p\right)_i$, that is, the first of inequalities (10.3) is replaced by

$$\lambda_S \mathcal{M}_i^{\max} + m_i = \left(M_p\right)_i. \tag{10.4}$$

Similarly, if the value of the hinge rotation at the section were negative, and had value $-\theta_i^-$, where θ_i^- is itself a positive quantity, then

$$\lambda_S \mathcal{M}_i^{\min} + m_i = -\left(M_p\right)_i. \tag{10.5}$$

If now equations (10.4) and (10.5) are multiplied through by the hinge rotation θ_i, then either

$$\left. \begin{array}{rcl} \lambda_S \mathcal{M}_i^{\max}\theta_i^+ + m_i\theta_i &=& \left(M_p\right)_i |\theta_i| \\ \text{or} \quad -\lambda_S \mathcal{M}_i^{\min}\theta_i^- + m_i\theta_i &=& \left(M_p\right)_i |\theta_i|, \end{array} \right\} \tag{10.6}$$

where the first equation is taken if θ_i is positive and the second if θ_i is negative. In either case the product $M_p\theta$ on the right-hand side is positive.

Equations (10.6) may now be summed for all hinges of the (virtual) mechanism, to give

$$\lambda_S \left[\Sigma \mathcal{M}_i^{\max}\theta_i^+ - \Sigma \mathcal{M}_i^{\min}\theta_i^- \right] + \Sigma m_i\theta_i = \Sigma \left(M_p\right)_i |\theta_i|. \tag{10.7}$$

Now m_i are self-stressing moments in equilibrium with zero external load, so that the sum $\Sigma m_i \theta_i$ is zero (cf. equation (2.9)). Thus the basic equation for incremental collapse is derived:

$$\lambda_S \left[\Sigma \mathcal{M}_i^{\max} \theta_i^+ - \Sigma \mathcal{M}_i^{\min} \theta_i^- \right] = \Sigma \left(M_p \right)_i |\theta_i|. \tag{10.8}$$

(It may be noted that the equation for static collapse, equation (2.19), may be recovered from equation (10.8). If the loads do not fluctuate between limits, but have fixed values, then $\mathcal{M}_i^{\max} = \mathcal{M}_i^{\min} = \mathcal{M}_i$, say, and equation (2.19) at once follows.)

The relationship between the shakedown load factor λ_S and the static load factor λ_C for maximum values of the same loads may be determined by further examination of equation (10.8), in a way that was first given by Ogle (1964). Suppose a unit load acting at section j of a frame produces an elastic bending moment μ_{ij} at section i of the frame. Then the actual load W_j acting at j will give rise to an elastic bending moment

$$\mathcal{M}_i = \mu_{ij} W_j. \tag{10.9}$$

The load W_j is confined within given limits, inequalities (10.1), and the value $W_j^{\max}$ or $W_j^{\min}$ will be chosen for the purpose of calculating $\mathcal{M}^{\max}$ according as μ_{ij} is positive or negative, say μ_{ij}^+ or $-\mu_{ij}^-$, where μ_{ij}^+ and μ_{ij}^- are themselves positive numbers. Thus

$$\left. \begin{array}{rcl} \mathcal{M}_i^{\max} & = & \Sigma_j \left(\mu_{ij}^+ W_j^{\max} - \mu_{ij}^- W_j^{\min} \right), \\ \text{and similarly} & & \\ \mathcal{M}_i^{\min} & = & \Sigma_j \left(-\mu_{ij}^- W_j^{\max} + \mu_{ij}^+ W_j^{\min} \right). \end{array} \right\} \tag{10.10}$$

The basic incremental collapse equation, equation (10.8), may now be written

$$\lambda_S \left[\Sigma_i \left\{ \Sigma_j \left(\mu_{ij}^+ W_j^{\max} - \mu_{ij}^- W_j^{\min} \right) \right\} \theta_i^+ \right. \\ \left. - \Sigma_i \left\{ \Sigma_j \left(-\mu_{ij}^- W_j^{\max} + \mu_{ij}^+ W_j^{\min} \right) \right\} \theta_i^- \right] = \Sigma \left(M_p \right)_i |\theta_i| \tag{10.11}$$

or, using the compact summation convention of the tensor calculus, in which summation is always carried out if a suffix is repeated within a product,

$$\lambda_S \left(\mu_{ij}^+ W_j^{\max} \theta_i^+ - \mu_{ij}^- W_j^{\min} \theta_i^+ + \mu_{ij}^- W_j^{\max} \theta_i^- - \mu_{ij}^+ W_j^{\min} \theta_i^- \right) \\ = \left(M_p \right)_i |\theta_i|. \tag{10.12}$$

The corresponding equation of static collapse, written for the condition that the loads have their maximum values, is

$$\lambda_C \left(\mu_{ij}^+ W_j^{\max} \theta_i^+ - \mu_{ij}^- W_j^{\max} \theta_i^+ + \mu_{ij}^- W_j^{\max} \theta_i^- - \mu_{ij}^+ W_j^{\max} \theta_i^- \right) \\ = \left(M_p \right)_i |\theta_i|. \tag{10.13}$$

The two equations are, of course, almost identical; the only difference is that all loads in equation (10.13) are $W_j^{\max}$. If the two equations are divided by their load factors, and subtracted, then

$$
\left(\frac{1}{\lambda_{\mathrm{S}}} - \frac{1}{\lambda_{\mathrm{C}}}\right)\left[(M_p)_i\,|\theta_i|\right] = \theta_i^+ \mu_{ij}^- \left(W_j^{\max} - W_j^{\min}\right) \\
+ \theta_i^- \mu_{ij}^+ \left(W_j^{\max} - W_j^{\min}\right). \tag{10.14}
$$

The right-hand side of equation (10.14) is thus a measure of the difference between the reciprocals of λ_{C} and λ_{S}. Three powerful conclusions may be drawn.

First, it is the *range* of loading $\left(W_j^{\max} - W_j^{\min}\right)$ which leads to the difference between λ_{S} and λ_{C}, and not the absolute values of the loads. Thus any load of fixed magnitude (e.g. a dead load) cannot affect any term in equation (10.14), and may be disregarded in calculating the *difference* between λ_{S} and λ_{C}. Dead loads will of course play their part in the evaluation of the static collapse load factor λ_{C}, but do not enter the shakedown analysis if it is done according to equation (10.14).

Second, the whole of the right-hand side of equation (10.14) is positive or zero. The range of loading is itself essentially positive or zero, and the products $\theta^+ \mu^-$ and $\theta^- \mu^+$ are all, by definition, positive. Thus, for any mechanism θ,

$$
\lambda_{\mathrm{S}} \leq \lambda_{\mathrm{C}}. \tag{10.15}
$$

The shakedown load factor cannot exceed the corresponding static collapse load factor.

Third, it is only products $\theta^+ \mu^-$ and $\theta^- \mu^+$ which occur in equation (10.14). The difference between λ_{S} and λ_{C} arises only from loads which would produce a *positive* elastic bending moment at a section where there is a *negative* hinge rotation, or which would produce a *negative* elastic bending moment at a section where there is a *positive* hinge rotation. Equation (10.14) may be simplified by a change of notation; the right-hand side may be written

$$
\theta_i^+ \mu_{ij}^- \overline{W}_j + \theta_i^- \mu_{ij}^+ \overline{W}_j,
$$

and if the summation over the repeated suffix j is carried out, the expression may be further shortened to

$$
\theta_i^+ \overline{\mathcal{M}_i^-} + \theta_i^- \overline{\mathcal{M}_i^+} \equiv \left|\theta_i \mathcal{M}_i^*\right|, \text{ say,}
$$

where $\overline{\mathcal{M}_i}$ is the range of elastic bending moment (with a vestigial sign

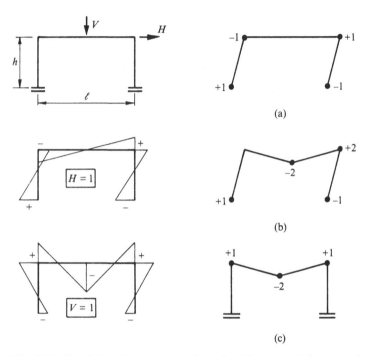

(a)

(b)

(c)

Fig. 10.1. Elastic bending moments for a fixed-base portal frame, and
possible collapse mechanisms.

attached) corresponding to the range of loading $\overline{W}_j$. It is seen that only the
products of a *positive* change of moment $\mathcal{M}_i^+$ with a *negative* hinge rotation
θ_i^-, and vice versa, are taken, and this is indicated by the final short notation
$|\theta_i \mathcal{M}_i^*|$.

Thus, finally, if the static collapse equation is written (and the summation
signs reintroduced)

$$\lambda_C \sum (M_w)_i \, \theta_i = \sum (M_p)_i \, |\theta_i| \qquad (10.16)$$

(cf. equation (2.16)), then the incremental collapse equation giving the
shakedown load factor may be written

$$\lambda_S \left[\sum (M_w)_i \, \theta_i + \sum |\mathcal{M}_i^* \theta_i| \right] = \sum (M_p)_i \, |\theta_i| . \qquad (10.17)$$

In these two equations, $(M_w)_i$ now represents *any* convenient set of equilibrium moments, and the results of the elastic analysis of the frame are
introduced only in the values of $\mathcal{M}^*$.

As an example, the collapse of an idealized fixed-base portal frame will be examined. Figure 10.1 shows the frame, of uniform section having full plastic moment M_p, acted upon by loads V and H. The values of V and H are supposed to vary randomly and independently within the ranges

$$\left. \begin{array}{l} 0 \leq V \leq V_0, \\ 0 \leq H \leq H_0. \end{array} \right\} \tag{10.18}$$

Also shown in fig. 10.1 are sketch elastic bending-moment diagrams for $H = 1$ and $V = 1$, together with the three possible modes of static or incremental collapse.

If the elastic solution for $H = 1$ is compared with the collapse diagram of mode (a), it will be seen that the signs of the bending moments at the hinge positions are the same in all cases as the signs of the hinge rotations. The conclusion is that there will be no '$\mathcal{M}^*$' terms arising from the side load H for mode (a), and that therefore the terms in H will be identical in the static collapse and the incremental collapse equation. On the other hand, the elastic solution for $V = 1$ indicates an opposition in sign for the hinges in the left-hand column for mode (a), so that there *will* be an '$\mathcal{M}^* \theta$' contribution arising from load V. The following equations may be derived:

$$\text{Mode (a)} \left\{ \begin{array}{llr} \text{Static:} & H_0 h & = 4M_p, \\[2mm] \text{Incremental:} & H_0 h + \dfrac{V_0 \ell}{8} \left(\dfrac{3\ell}{2\ell + h} \right) & = 4M_p. \end{array} \right\} \tag{10.19}$$

$$\text{Mode (b)} \left\{ \begin{array}{llr} \text{Static:} & H_0 h + \dfrac{V_0 \ell}{2} & = 6M_p, \\[2mm] \text{Incremental:} & H_0 h + \dfrac{V_0 \ell}{8} \left(\dfrac{9\ell + 4h}{2\ell + h} \right) & = 6M_p. \end{array} \right\} \tag{10.20}$$

$$\text{Mode (c)} \left\{ \begin{array}{llr} \text{Static:} & \dfrac{V_0 \ell}{2} & = 4M_p, \\[2mm] \text{Incremental:} & \dfrac{H_0 h}{2} \left(\dfrac{3h}{\ell + 6h} \right) + \dfrac{V_0 \ell}{2} & = 4M_p. \end{array} \right\} \tag{10.21}$$

Equations (10.19) to (10.21) are plotted in the interaction diagram of fig. 10.2 (this diagram has been drawn for $\ell = 2h$, but the general features will be preserved for other ratios of ℓ/h). It will be noted that the 'yield surface' for incremental collapse lies entirely within the corresponding yield surface for static collapse, as it must from inequality (10.15). That inequality was derived on the assumption that the *same* mechanism operated for both static and

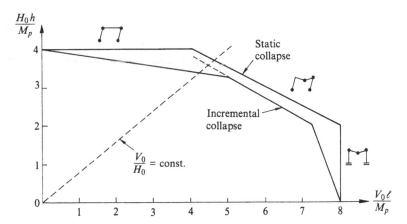

Fig. 10.2. Interaction diagram for fixed-base portal frame, showing loading limits for static and for incremental collapse.

incremental collapse. In fact, it is seen from the example of fig. 10.2 that an even more severe condition may be imposed in the analysis for incremental collapse; for a given ratio of V_0/H_0 (represented by a line through the origin), it is possible that the modes of collapse may differ in the static and incremental analyses.

The bending of a beam of trapezoidal cross-section

Section 7 of Tredgold's *Practical essay on the strength of cast iron* (1824) deals with the strength and deflexion of cast iron when it resists pressure or weight — that is, it deals with the elastic bending of beams. In the 1822 edition Tredgold considered only beams of cross-section having two axes of symmetry, but in the second edition of 1824 he added a paragraph (85a) – together with a footnote – which investigates a problem of bending with only one axis of symmetry:

> 85a. Hitherto we have only considered those forms where the neutral axis divides the section into identical figures; but there are some interesting cases[a] where this does not happen, such, for example, as the triangular section.
>
> [a] (They are interesting, because the earlier theorists fell into some serious errors respecting them, and consequently have led practical engineers into erroneous opinions.)

The triangular section considered by Tredgold is shown in fig. 11.1, with part of the triangle cut off at the vertex; this is a general case which will include that of the entire triangle (i.e. $m = 0$). Tredgold's immediate problem is to locate the neutral axis MN, and he does not have Navier's formal statement of 1826 that the neutral axis must pass through the centre of gravity of the cross-section.

Navier (1826) was in fact the first to make this statement, but the result is implicit in earlier work, for example that of Parent (1713) and of Coulomb (1773). Both Parent and Coulomb were aware that the net longitudinal force on a cross-section in pure bending must be zero. This requirement of statical equilibrium, coupled with the idea of a linear distribution of strain, and hence of stress for a linear-elastic material, leads at once to the correct position of the neutral axis.

Tredgold was not aware of this requirement. Instead, he stated that the

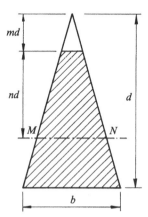

Fig. 11.1. A trapezoidal cross-section derived from an isosceles triangle.

two portions of the cross-section, one above the neutral axis and one below, must have the same 'strength'. Thus the second moment of area about MN of the portion of the cross-section lying above MN, divided by the ordinate nd, must equal the equivalent expression obtained for the portion lying below MN. Tredgold derives an equation expressing this equality:

$$n^2(4m + n) = 4(1 - m - n)^2 - (1 - m - n)^3.$$

This equation is actually a quadratic in n, and Tredgold obtains the general solution.

If the cross-section has two axes of symmetry then Tredgold's procedure leads to the correct answer, but the results are in error for the trapezoidal section of fig. 11.1. For example, for the complete triangle for which $m = 0$, Tredgold obtains $n = 0.697$ instead of the true value $2/3$, and he deduces that the section modulus for the full triangle is $0.0565\ bd^2$ instead of $bd^2/24$. Tredgold has a footnote to the effect that this result was first given in the *Philosophical Magazine* of 1816.

Indeed, 'T.T.' gives 'Rules for ascertaining the Strength of Materials' on p.22 of the *Philosophical Magazine* for that year; 'T.T.' states that the results are new, and he gives correct values for the section moduli of beams of rectangular cross-section, for a square bent about a diagonal, for a circle and for a tube. Finally, for the triangle, T.T.'s result is incorrect, and is given as $0.05643\ bd^2$. This is very slightly different from the value given by Tredgold eight years later — in any case, it is this later version which is referred to by Saint-Venant, in his 1864 edition of Navier (1826). Saint-Venant ascribes the *principe singulier* (of equal 'strengths' above and below the neutral axis) to Tredgold; as has been noted, Tredgold's principle happens to give the correct result for doubly symmetric sections.

However, Tredgold is interested in the curious behaviour of his (incorrect) expression for the section modulus of the trapezium. As the value of *m* is increased from zero, so that the cross-section is reduced, so the value of the section modulus increases. Tredgold deduces numerically that the section will have a greatest strength when $m = 0.1$, at which value the modulus is $0.058 \, bd^2$, or about 3 per cent greater than the (incorrect) value for the entire triangle. He notes that William Emerson first announced this seeming paradox in 1754.

The 1825 edition of Emerson's *The principles of mechanics* discusses the problem of the bending of a beam of arbitrary cross-section, and in particular he deals with the section of fig. 11.1. However (as remarked by Tredgold) Emerson places his neutral axis at the base of the cross-section, and adopts (in 1825!) what is effectively the approach of Mariotte (1686). From this faulty analysis he deduces that if one-ninth of the triangular prism is cut away (i.e. $m = 1/9$ in fig. 11.1), ' ... the remaining beam will bear a greater weight ... than the whole ..., or the part will be stronger than the whole, which is a paradox in mechanics.'

Despite all these mistakes, the paradox is genuine. The correct value of *n*, locating the neutral axis, is given by

$$n = \frac{(1-m)(2+m)}{3(1+m)},$$

and the general value of the section modulus is

$$\frac{1}{12}bd^2 \left[\frac{(1-m)^2(1+4m+m^2)}{(2+m)} \right].$$

This section modulus is a maximum when

$$m^3 + 5m^2 + 7m - 1 = 0,$$

that is, for $m = 0.1304$, when the modulus has value

$$\frac{1}{12}bd^2(0.5461) = 0.0455 \, bd^2.$$

Thus the removal of about one-eighth of the depth of the triangle results in an increase of elastic strength of about 9 per cent.

The paradox arises, of course, because the 'strength' of the cross-section is determined by the attainment of a limiting stress at a single point. This might be a fair measure of strength if the material were brittle, and liable to fracture in the presence of such a limiting stress — the material would not then, however, be suitable for general structural use. A ductile material will exhibit some plastic behaviour, and if the 'true' strength of the cross-section is assessed by a calculation of the full plastic modulus, then common sense is not offended. The plastic modulus has value

$$\frac{1}{3}bd^2 \left[\left(1 + m^3\right) - \frac{1}{\sqrt{2}} \left(1 + m^2\right)^{\frac{3}{2}} \right],$$

and this has its largest value, $0.0976\, bd^2$, for $m = 0$.

The simple plastic bending of beams

As will have been noted from the last chapter, the theory of bending of beams seems always to have given some difficulty. The first key requirement of statics, that there should be no net thrust across a cross-section in pure bending, was recognized in the eighteenth century; but it was only in 1826 that Navier stated explicitly that as a consequence the neutral axis must pass through the centre of gravity of the cross-section. However, even Navier was not aware of the consequences of a second statical requirement; moments of the forces acting on a cross-section lead to the notion of principal axes of bending. Thus Navier gave wrong expressions for the bending of a rectangular cross-section about an axis not parallel to one of its sides, and it fell to Saint-Venant in his 1864 edition of Navier to discuss fully the question of principal second moments of area.

Saint-Venant extended his analysis to cover non-linear behaviour of the material, but confined his work in this connexion to symmetrical cross-sections. The elastic/perfectly plastic material is a special case of Saint-Venant's more general material, and the plastic bending problem was considered separately by Ewing (1899). Ewing again discussed only the rectangular section bent about a principal axis, and indeed most of the modern standard texts on plastic theory do not treat the unsymmetrical problem. Brown (1967) seems to be the first to have recorded the general features of plastic unsymmetrical bending. He gives no specific solutions, but notes that the principal axes of elastic and plastic bending need not coincide (this is a property also of sections having only one axis of symmetry), and that the principal axes of plastic bending are not necessarily orthogonal.

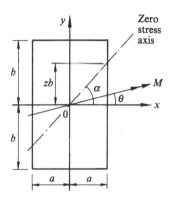

Fig. 12.1. A rectangular cross-section bent about an inclined axis.

12.1 The rectangular cross-section

Figure 12.1 shows a rectangular cross-section subjected to a bending moment M which acts about an axis included at an angle θ to the coordinate axis $0x$ (which is of course one of the principal axes of elastic bending). The cross-section is supposed to be fully plastic, so that all material above and to the left of the zero-stress axis is at the yield stress σ_0 (in tension, say), while the material below and to the right of the zero-stress axis is at the same stress σ_0 (in compression). In elastic bending the neutral axis and the axis of the applied bending moment do not necessarily coincide; similarly, the zero-stress axis for fully plastic bending makes an angle α in fig. 12.1, which will in general differ from the angle θ.

The parameter z is marked in fig. 12.1, where

$$\tan \alpha = \frac{b}{a} z, \tag{12.1}$$

and, by taking moments about $0x$ and $0y$ in turn, the components M_x and M_y of the bending moment M are found to be

$$\left. \begin{array}{l} M_x = M \cos \theta = 2ab^2\sigma_0 \left(1 - \tfrac{1}{3}z^2\right), \\ M_y = M \sin \theta = 2a^2b\sigma_0 \left(\tfrac{2}{3}z\right). \end{array} \right\} \tag{12.2}$$

These expressions hold for $-1 \leq z \leq 1$, that is, for values of $\tan \alpha$ less than b/a; similar expressions may be derived by interchange of the symbols when $\tan \alpha$ is greater than b/a.

Equations (12.2) are of course parametric expressions for the yield surface

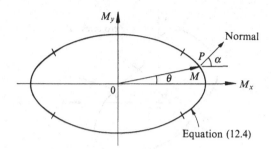

Fig. 12.2. Yield surface for the rectangular cross-section.

for the rectangular cross-section, and can be rearranged to give

$$\left.\begin{array}{l} M^2 = M_x^2 + M_y^2, \\[2mm] \tan\theta = \dfrac{M_y}{M_x}, \end{array}\right\}$$ (12.3)

and

$$\left(\frac{M_x}{2ab^2\sigma_0}\right) + \frac{3}{4}\left(\frac{M_y}{2a^2b\sigma_0}\right)^2 = 1.$$ (12.4)

Equation (12.4) is sketched in fig. 12.2, together with a similar equation for $\tan\alpha$ greater than b/a, and both of these with the signs reversed (bending in the opposite sense). Equations (12.2) show that the vector OP to the general point P on the curve represents the bending moment M.

The value of α marked in fig. 12.2 is the same as that marked in fig. 12.1; if the value of $-dM_x/dM_y$ is calculated from equation (12.4) for the yield surface, then equation (12.1) for $\tan\alpha$ is recovered, and the normality rule of plasticity has been confirmed. Further, it will be seen from fig. 12.2 that the values of θ and α are equal for $\theta = 0$ and $\pi/2$; the principal axes for plastic bending of the rectangular cross-section are, as might be expected, the axes of symmetry.

12.2 The general unsymmetrical section

An arbitrary reference frame of axes is shown for the general cross-section of fig. 12.3; a bending moment M is applied about an axis making an angle θ with $0x$. If the cross-section is fully plastic, then the zero-stress axis dividing the zone yielding in tension from that yielding in compression will not in general coincide with the axis of the bending moment (just as for the case of

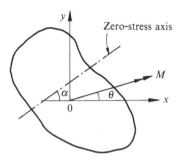

Fig. 12.3. General cross-section.

the rectangular cross-section). The condition of no net axial thrust requires the zero-stress axis to divide the cross-section into two equal areas. As the axis of the applied bending moment changes the zero-stress axis will shift; it will remain an equal-area axis but, for the general asymmetrical cross-section, it will not pass through a fixed point.

Just as for the rectangular section, the components M_x and M_y of the bending moment M may be used as coordinate axes to plot a yield surface. This yield surface must be skew-symmetric, since the numerical values of M_x and M_y will be the same if the angle θ in fig. 12.3 is increased by π, although the signs will be reversed. A typical yield surface of this kind is sketched in fig. 12.4. The vector M to a general point P makes an angle θ with the reference direction, while the normal to the curve makes an angle α corresponding with the angle of the zero-stress axis. There will be in general two values of θ for which $\alpha = \theta$, that is, for which the axis of deformation is parallel to the axis of the applied bending moment; these directions define the plastic principal axes. The axes are indicated in fig. 12.4; they are located by the points of tangency of the inscribed and escribed circles centred on the origin.

As sketched, the plastic principal axes are clearly not orthogonal — nor need there be any coincidence with an elastic principal axis. The elastic orthogonality condition, not explored by Navier, arises from the assumed linear stress distribution across the section, whatever its shape. In fig. 12.5 an elastic principal axis is shown passing through G, the centre of gravity of the cross-section. If this neutral axis is taken also to be the axis of x, then the bending moment about the axis Gy at right angles must be zero. This leads at once to the equation

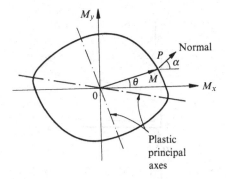

Fig. 12.4. Skew-symmetrical yield surface for the general cross-section.

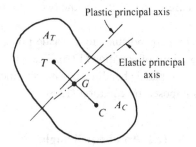

Fig. 12.5. Elastic and plastic principal axes of bending will differ for the general cross-section.

$$\int^A xy\,dA = 0,\qquad\qquad (12.5)$$

where dA is an element of cross-sectional area, and the integration covers the whole cross-section. Equation (12.5) characterizes the elastic principal axes; since the equation is symmetrical in x and y, the directions of these principal axes will be orthogonal.

By contrast, the plastic principal axis sketched in fig. 12.5 does not generally pass through G, but divides the cross-section into two equal areas A_C and A_T, whose centres are at C and T respectively. Evidently CGT is a straight line, and $CG = GT$. If the plastic principal axis is now taken to be the axis of x, then the condition that the axis is indeed principal implies that, as before, the bending moment about the y axis must be zero, that is

$$\int^{A_T} x\,dA = \int^{A_C} x\,dA,\qquad\qquad (12.6)$$

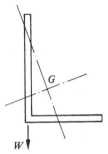

Fig. 12.6. Elastic principal axes for the angle section.

where the integrations are carried out over the two separate areas, in tension and in compression.

The line CT must be perpendicular to a plastic principal axis, and there are, in general, two solutions of equation (12.6). There will be a "strong" and a "weak" principal axis of plastic bending, but no condition of orthogonality between these axes is imposed by equation (12.6).

12.3 An unequal angle

The unequal angle has no axes of symmetry. Elastic principal axes are sketched in fig. 12.6, and elastic behaviour must be examined by resolving the applied bending moment into components applied about these axes. If, for example, the angle section were used as a cantilever with one leg vertical, as indicated in fig. 12.6, then the tip load W must be resolved into components parallel to the directions of the principal axes. If the resulting deflexions are superimposed, it is seen that the tip of the cantilever will move both horizontally and vertically under the action of a purely vertical load.

Similar behaviour occurs when the section becomes fully plastic. The plastic principal axes do not coincide with the corresponding elastic axes, but they are certainly not horizontal and vertical. Thus the cantilever beam of fig. 12.6 will be bent about an axis which is not parallel to a plastic principal axis; at collapse of the beam, the tip will again move both horizontally and vertically, although the direction of motion may differ markedly from that of the elastic cantilever.

The skew-symmetric yield surface for the unequal angle will be calculated for an idealized section composed of two thin rectangles. In fig. 12.7 the legs

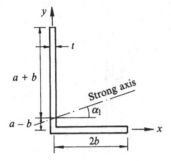

Fig. 12.7. Angle in fully plastic state bent about the 'strong' axis.

have lengths $2a$ and $2b$ and equal small thickness t; the assumption that t/a is small will introduce small errors if centre-line dimensions of real angles are substituted in the formulae derived below.

The area of the whole cross-section is $2(a+b)t$, and the zero-stress axis at full plasticity will divide the cross-section into two equal parts, each of area $(a+b)t$. Figure 12.7 shows the case where the zero-stress axis lies in the 'strong' principal direction, making an angle α_1 with the x-axis. Geometrically, the centre of tension T and the centre of compression C (cf. fig. 12.5) could be found for the areas lying on either side of the zero-stress axis; the line CT, perpendicular to the plastic principal axis, would then give the value of α_1. In fact, the bending moments M_x and M_y acting about the x and y axes may be computed for the fully plastic state of fig. 12.7, and then combined to give the required condition.

If the yield stress of the material is σ_0, then the values of the bending moments are

$$\left. \begin{aligned} M_x &= \left(a^2 + 2ab - b^2\right) t\sigma_0, \\ M_y &= 2b^2 t\sigma_0. \end{aligned} \right\} \tag{12.7}$$

It will be seen that these expressions are independent of the inclination of the zero-stress axis; indeed, from fig. 12.7, the value of α_1 can lie between $-\tan^{-1}(a-b)/(2b)$ and $+\frac{1}{2}\pi$. However, the inclination θ of the axis of the applied bending moment is fixed from equations (12.7) by

$$\tan\theta = \frac{M_y}{M_x} = \frac{2b^2}{a^2 + 2ab - b^2}. \tag{12.8}$$

Evidently equations (12.7) correspond to a corner on the yield surface (in

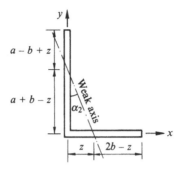

Fig. 12.8. Angle in fully plastic state bent about the 'weak' axis.

fact, they correspond to point A in fig. 12.9, as will be seen). The value of θ given by equation (12.8) is equal to that of α_1.

Table 12.1 gives some numerical results, using centre-line dimensions and 'book' values from standard section tables. It will be seen that there are only small differences between the calculated directions of the strong plastic principal axes and the corresponding elastic values from the tables.

Table 12.1.

	a(mm)	b(mm)	$\tan\alpha_1$	α_1	Elastic
Equal angle $(a = b)$	—	—	1	45°	45°
100 × 75 × 12 × 15.4 kg/m	47	34.5	0.559	29.2°	28.4°
150 × 75 × 12 × 20.2 kg/m	72	34.5	0.266	14.9°	14.5°

The 'weak' plastic principal axis cuts both legs of the angle. In fig. 12.8 its location is specified by the parameter z, whose value is to be determined; the axis has been drawn as an equal area axis, cutting the section into two halves each of area $(a + b)t$. The inclination of the axis is given by

$$\tan\alpha_2 = \frac{z}{a + b - z}. \tag{12.9}$$

A second expression for $\tan\alpha_2$ may be found from the condition that the axis of the applied bending moment is parallel to the plastic principal axis.

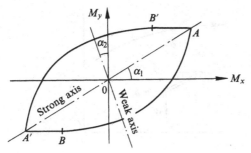

Fig. 12.9. Yield surface for the unequal angle.

For the position of the axis shown in fig. 12.8,

$$M_x = \left\{ \left(a^2 - 2ab - b^2 \right) + 2\left(a + b \right) z - z^2 \right\} t\sigma_0, \atop M_y = \left(z^2 - 2b^2 \right) t\sigma_0. \right\} \quad (12.10)$$

Thus

$$\tan \alpha_2 = \frac{\left(a^2 - 2ab - b^2 \right) + 2\left(a + b \right) z - z^2}{2b^2 - z^2}. \quad (12.11)$$

The simultaneous solution of equations (12.9) and (12.11) gives the value of z and hence of α_2; table 12.2 gives some numerical results.

Table 12.2.

	z(mm)	$\tan \alpha_2$	α_2	Elastic
Equal angle($a = b$)	a	1	45°	45°
100 × 75 × 12 × 15.4 kg/m	20	0.327	18.1°	28.4°
150 × 75 × 12 × 20.2 kg/m	5	0.052	3.0°	14.5°

The direction of the weak plastic axis is somewhat different from that of the corresponding elastic principal axis for both the unsymmetrical sections. Further, the strong and weak axes are not orthogonal, but intersect at under 80° for both sections.

The yield surface may be plotted for the cross-section, and the above analysis gives the information necessary to discuss plastic bending about any axis, not necessarily a principal plastic axis. In fig. 12.9, equations (12.10) plot to give the curve AB, and with signs reversed (bending in the opposite sense),

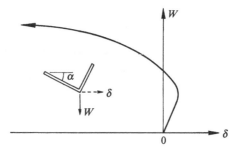

Fig. 12.10. The angle tested as a cantilever: schematic load/deflexion curve.

they give the curve $A'B'$. At the point B the value of z (fig. 12.8) is zero, and for the portions BA' and $B'A$ of the yield surface the zero-stress axis lies wholly within the longer leg of the angle section. For the approximation of very thin rectangles this means that the value of M_x is indeterminate, and BA' and $B'A$ are straight lines parallel to the axis of M_x. As noted above, equations (12.7) plot as the single point A.

The general shape of fig. 12.9 will be preserved for any angle section; the particular proportions of the figure correspond to a ratio $a/b = 4/3$. The strong and weak axes are marked; the latter is, of course, normal to the yield surface. The normality rule applies in general (cf. the yield surface of fig. 12.4); for bending about an axis which gives components M_x, M_y lying on $B'A$ in fig. 12.9, for example, deformation of the section will occur about an axis parallel to the longer leg of the angle.

The difference in inclination between the weak plastic axis and the corresponding elastic principal axis is large enough to be observed experimentally; this difference is over $10°$ for the 100×75 angle; see table 12.2. If this angle were mounted as a cantilever carrying a tip load, and orientated so that bending occurs about an axis close to the weak axis, then both vertical and lateral movement of the tip would occur, in general. If, however, the angle were mounted so that the value of α in the inset figure in fig. 12.10 were exactly $28.4°$, then no lateral movement should occur so long as the material remained elastic. As the full plastic moment was approached the axis of bending would gradually shift towards its final position, given by $\alpha_2 = 18.1°$ in table 12.2, and this shift would be accompanied by lateral deflexion of the tip of the cantilever.

If, on the other hand, the test were arranged so that the value of α in fig. 12.10 lay somewhere between the $18.1°$ and $28.4°$ of table 12.2, then

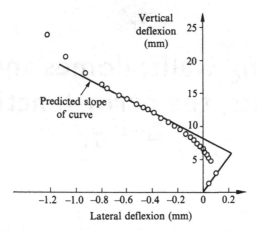

Fig. 12.11. Results from a test on a cantilevered angle.

the response of the cantilever would be an initial lateral deflexion in one direction, followed by a reversal of this direction as the material became plastic. Figure 12.10 shows schematically the results corresponding to an actual test.

The results themselves are shown in a different way in fig. 12.11, in which vertical deflexion is plotted against lateral deflexion of the tip of the cantilever. The predicted slope of the curve in the plastic region is obtained from the direction of the normal at the appropriate point of the yield surface of fig. 12.9.

Leaning walls; domes and fan vaults; the error function $\int e^{-t^2} dt$

Masonry may be regarded as an assemblage of dry stones (or bricks or other similar material), some squared and fitted and some not, placed together to form a stable structure. Any mortar that may have been used will have been weak, and may have decayed with time; it cannot be assumed to add strength to the construction. Stability of the whole is assured, in fact, by the compaction under gravity of the various elements; a general state of compressive stress can exist, but only feeble tensions can be resisted.

This is the unilateral model of masonry; the material can resist compression, but has zero tensile strength. Further, compressive stresses in masonry structures are usually very low indeed, so that the material may be assumed, effectively, to be infinitely strong in compression. To add a final and imprecise assumption, it is clear that individual components of a masonry structure must in fact possess tensile strength, even if the overall assemblage has none. As an example, it is easy to envisage a dry stone wall in which the stones can indeed be lifted away, but which, in the absence of such interference, will retain its structural shape. The stones must, however, have a certain shape and size; an attempt to build a vertically sided wall from small particles (sand) will be unsuccessful.

13.1 The leaning wall

The foundations of a wall will be supposed to give way differentially, so that the wall tilts. A monolithic rectangular block, of height H and width b, may be tilted on its base until its centre of gravity is vertically above one corner; slight further movement will cause the block to topple. The cosine of the

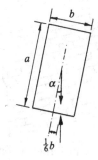

Fig. 13.1. A tilted block of masonry.

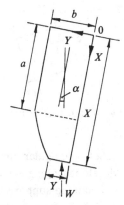

Fig. 13.2. A tilted block for which the support force falls outside the
middle third of the original width. A fracture has developed
whose shape is to be determined.

angle of tilt will be approximately unity for a block whose height to width
ratio is 4 or more, so that the critical displacement, the 'lean' of the block, is
equal to its width. The calculations are not so simple for unilateral masonry.

Figure 13.1 shows a rectangular block of masonry of height a tilted to
such an angle α that the support force acts just at the limit of the middle
third of the section. At this condition, according to simple elastic theory,
the left-hand bottom corner will be just free of stress, and the block will be
supported by linearly increasing compressive forces along the bottom surface.
Evidently

$$\tan \alpha = \frac{1}{3}\frac{b}{a}. \tag{13.1}$$

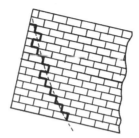

Fig. 13.3. An 'actual' fracture (cf. fig.13.2).

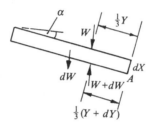

Fig. 13.4. A thin slice of masonry.

Figure 13.2 shows part of a taller rectangular block at the same inclination α, for which a fracture has developed along a stress-free surface; some of the masonry has fallen away, remaining supported by the tilted foundation, as shown schematically in fig. 13.3. A slight further inclination will transfer some of the 'bricks' just above the fissure to the passive pile below.

The shape of the free surface in fig. 13.2 is to be determined. At the general section, distant X from the origin, the total weight W of the masonry is supported by a force again acting at the limit of the middle third of a base of reduced dimension Y. Figure 13.4 shows an elemental slice of the wall, and, for unit weight of material, $dW = Y\,dX$, or

$$Y = \frac{dW}{dX}. \tag{13.2}$$

By taking moments about A, an equilibrium equation may be established:

$$\frac{1}{6}Y\frac{dW}{dX} - W\tan\alpha = \frac{1}{3}W\frac{dY}{dX}. \tag{13.3}$$

Using equation (13.2) to eliminate X,

$$\frac{1}{6}Y^2 - W\tan\alpha = \frac{1}{3}WY\frac{dY}{dW}. \tag{13.4}$$

The non-dimensional variables

$$x = \frac{X}{a}, \quad y = \frac{Y}{b}, \quad w = \frac{W}{ab}$$

may be introduced, and, using equation (13.1), equation (13.4) becomes

$$y^2 = 2w \left(1 + y \frac{dy}{dw} \right), \tag{13.5}$$

and this has for solution

$$y^2 = w \left(C - 2 \log w \right), \tag{13.6}$$

where C is a constant of integration. In fig. 13.2 the crack starts at $X = a$, $Y = b$, at which condition $W = ab$, so that equation (13.6) must satisfy the condition $x = y = w = 1$. Thus $C = 1$. From equation (13.2), $y = dw/dx$, so that equation (13.6) becomes

$$\frac{dw}{dx} = \sqrt{w \left(1 - 2 \log w \right)}, \tag{13.7}$$

that is,

$$x = D + \int_1^w \frac{dw}{\sqrt{w \left(1 - 2 \log w \right)}}, \tag{13.8}$$

where D is a second constant of integration whose value is determined as unity from the condition $x = w = 1$.

Equation (13.8) may be solved in terms of a parameter t, where

$$1 - 2 \log w = 4t^2; \tag{13.9}$$

making this substitution,

$$x = 1 + 2e^{1/4} \int_t^{1/2} e^{-t^2} dt. \tag{13.10}$$

The integral in equation (13.10) is (with the factor $2/\sqrt{\pi}$) that of the error function erf t. The final results may be collected together:

$$\left. \begin{array}{l} w = e^{1/2} e^{-2t^2} \\ x = 1 + \sqrt{\pi} e^{1/4} \left(\text{erf} \ \tfrac{1}{2} - \text{erf} \ t \right) \\ y = 2e^{1/4} t e^{-t^2}. \end{array} \right\} \tag{13.11}$$

The shape of the fissure corresponding to equations (13.11) is plotted in fig. 13.5; this non-dimensional sketch for the parameters x and y is plotted for $a = b$, and may be stretched linearly to correspond to the parameters X and Y for the real block of masonry.

Fig. 13.5. An accurate non-dimensional plot of the shape of a fracture
of a masonry wall on the point of collapse, drawn for $a = b$
(see fig.13.2). The figure may be 'stretched' to give the shape
for any ratio a/b.

The maximum height of the masonry block, for the given lean α, is given
by the condition that the crack penetrates through the whole width, that
is by the condition $y = 0$, for which the parameter t is also zero. The
dimensionless height h of a wall that is just becoming unstable is given, from
the second of equations (13.11), by

$$h = 1 + \sqrt{\pi}e^{1/4}\left(\text{erf } \tfrac{1}{2}\right),$$

and since

$$\text{erf } \tfrac{1}{2} = 0.5205,$$

$$h = 2.1846.$$

Thus the height H of the actual wall is related to the width b by

$$\frac{H}{b} = \frac{ah}{b} = \left(\frac{1}{3}\cot\alpha\right)(2.1846),$$

that is

$$\tan\alpha = \frac{0.7282}{(H/b)}. \tag{13.12}$$

At this condition, the out-of-plumbness of the top of the tower with respect
to its base, that is its 'lean', is $H\sin\alpha$ and, since $\sin\alpha \simeq \tan\alpha$, the maximum
lean is $0.728b$ (compared with b had the wall been monolithic).

These solutions may be applied to the discussion of the problem of leaning
towers, of which perhaps the most famous is that of Pisa. There are, however,
many other examples of such towers, both in Italy (in Venice and the islands
of the lagoon) and in other parts of the world.

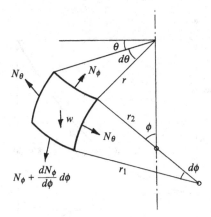

Fig. 13.6. Stress resultants acting on an element of a domical shell.

13.2 The masonry dome

The masonry dome will first be considered as a thin shell, and conventional membrane theory will be used to evaluate the stresses in the dome due to self-weight. As will be seen, this approach must finally be abandoned for the real dome made of unilateral material.

An element of a shell of revolution is shown in fig. 13.6; the generating curve has a radius of curvature r_1. The shell is symmetrical and of uniform thickness with self weight w per unit area; under these conditions there are two independent stress resultants, N_ϕ in the meridional direction and N_θ in the hoop direction. Equilibrium of the element demands that

$$\frac{d}{d\phi}(rN_\phi) - r_1 N_\theta \cos\phi + wrr_1 \sin\phi = 0 \qquad (13.13)$$

and

$$\frac{N_\phi}{r_1} + \frac{N_\theta}{r_2} = -w\cos\phi \qquad (13.14)$$

where

$$\left.\begin{aligned} r &= r_2 \sin\phi \\ \text{and} \quad r_1 \cos\phi &= \frac{dr}{d\phi}. \end{aligned}\right\} \qquad (13.15)$$

The solution of these equations for the spherical dome ($r_1 = r_2 = a$) is

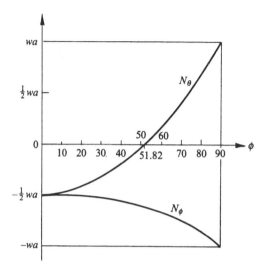

Fig. 13.7. Stress resultants for a hemispherical dome under its own
weight, according to simple shell theory.

particularly simple, and the stress resultants are given by

$$N_\phi = -\frac{wr}{1 + \cos \phi}$$
$$\text{and} \quad N_\theta = -wr \cos \phi - N_\phi. \qquad (13.16)$$

These expressions are sketched in fig. 13.7, and it will be seen that while the
meridional stress remains compressive throughout the quadrant, the hoop
stress is compressive at the crown but changes sign and becomes tensile at an
angular distance of 51.82° from the crown (the solution of $\cos^2 \phi + \cos \phi = 1$,
that is, $\cos \phi = \frac{1}{2} \left(\sqrt{5} - 1 \right)$).

Such tensile stresses are inadmissible for the unilateral masonry structure.
If a hemispherical dome is to behave as dictated by the membrane analysis,
then the tensile behaviour implied by the sketch of fig. 13.8 must be induced
by some other means. For example, encircling chains could be used in an
attempt to prevent the 'bursting' of the dome.

In practice, the slight outward spread of the base of the dome, induced by
the outward thrust of the masonry, will lead at once to meridional cracking,
rising from the base and dying out twoards the crown. This was the state
of the dome of St Peter's, Rome, on which Poleni reported in 1748, about
two centuries after the completion of Michelangelo's design (the dome was

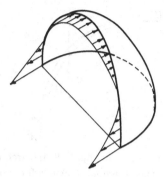

Fig. 13.8. Hoop stresses N_θ of fig. 13.7 acting on the meridian of a hemispherical dome.

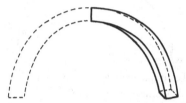

Fig. 13.9. A quasi two-dimensional arch formed by slicing a hemispherical dome along neighbouring meridians.

actually built by the engineers Fontana and Della Porta). Poleni noted that the cracks had already divided the dome into portions approximating half spherical lunes (orange slices); for the purpose of his analysis, he sliced the dome hypothetically into 50 such lunes. He then considered the stability of, effectively, a two-dimensional arch composed of two lunes leaning against each other at the crown, as shown in fig. 13.9. If such an arch were in itself stable, then he argued, correctly, that the whole dome, cracked or not, would also be stable.

In order to confirm the stability of the arch, Poleni made use of Hooke's theorem of 1676: 'As hangs the flexible line, so but inverted will stand the rigid arch'; that is, the shape of a tensile hanging chain is identical with that of a compressive line of thrust to carry the same loads. If this line lies within the material of the arch, then stability will be assured. From a drawing of the cross-section of the dome of St Peter's, Poleni calculated the weight of a lune, dividing it into 16 segments for this purpose and making due allowance for the weight of the lantern. He then loaded a flexible string

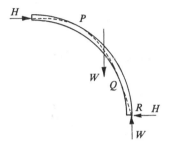

Fig. 13.10. The line of thrust for fig. 13.9; the arch is drawn to scale to show the least possible thickness for stability.

Fig. 13.11. Collapse mechanism for the arch of fig. 13.9 having the least thickness of fig. 13.10.

with 32 weights, each weight in proportion to a segment of the two lunes, and so determined experimentally the shape of the hanging chain. This shape indeed lay between the surfaces of the masonry.

Poleni had established a solution for the dome for which the hoop stress is zero. He had found a way in which a spherical dome could satisfy the conditions of equilibrium, but in which the forces are no longer compelled to lie in a spherical surface (as they are by the simple membrane analysis leading to equations (13.16)). Rather the forces follow the line of the 'hanging chain', and a certain thickness of masonry is necessary to contain this line. It is of interest to examine the minimum thickness of spherical lunes which will just be stable.

Figure 13.10 shows half of the two-lune arch of such a thickness that it can only just contain the thrust line; hinges are just forming at sections P, Q and R, and the corresponding collapse state for the complete dome is sketched in fig. 13.11. Portion PP of the dome near the crown remains coherent, while the hinges at P, Q and R ensure separation of the lunes so that no hoop stress can be generated. If the arch of fig. 13.10 did not embrace a full semicircle but had a cut-off angle ϕ_0 less than $90°$, then the thickness of the

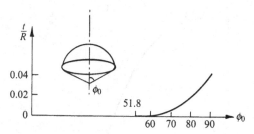

Fig. 13.12. Least thicknesses for incomplete spherical domes.

arch could be reduced, and fig. 13.12 plots the required minimum thickness t/R of the dome as a function of ϕ_0. For a full hemisphere the minimum thickness is just over 4 per cent of the radius. At $\phi_0 = 51.82°$ points P and Q coincide and the thickness falls to zero; for cut-off angles less than this a spherical dome can be built, in theory, with a vanishingly small thickness (cf. fig. 13.7).

The results quoted above may be obtained from the statics of fig. 13.10; equally, they may be obtained (in a different form) from the membrane equations (13.13), (13.14) and (13.15). Solutions are sought for which the hoop stress N_θ is zero, but the dome is no longer assumed to be spherical; rather, the radius of curvature r, defining the generating curve, is allowed to be a variable. The equations combine at once to give

$$\frac{d}{d\phi}(wrr_1 \cos \phi) = wrr_1 \sin \phi,$$

that is $wrr_1 \cos^2 \phi = \text{const.}$

Using the second of equations (13.15),

$$wr\frac{dr}{d\phi}\cos \phi = \text{const} = \frac{x_0^2}{2} \text{ (say)}. \tag{13.17}$$

Equation (13.17) integrates at once to give

$$r^2 = x_0^2 \log(\sec\phi + \tan \phi), \tag{13.18}$$

where a second constant of integration is zero for a dome horizontal at its crown ($\phi = 0$ at $r = 0$).

The rectangular coordinate system of fig. 13.13 may be introduced to effect the final integration; the radius r is identified with the abscissa x, and $\tan \phi = dy/dx$. From equation 13.18,

$$\sec \phi + \tan \phi = e^{x^2/x_0^2}, \tag{13.19}$$

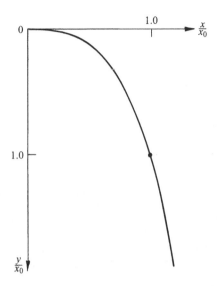

Fig. 13.13. The 'perfect' shape of dome in which the form follows the
line of thrust; the curve in the figure (for the portion
between (0,0) and (1,1)) can be fitted within the circles of
fig. 13.10.

whence

$$\tan \phi = \frac{dy}{dx} = \frac{1}{2} \left(e^{x^2/x_0^2} - e^{-x^2/x_0^2} \right). \tag{13.20}$$

Hence, using the substitution $t = x/x_0$,

$$y = \frac{1}{2} x_0 \left[\int_0^{x/x_0} e^{t^2} dt - \int_0^{x/x_0} e^{-t^2} dt \right]. \tag{13.21}$$

The error integral has again appeared; the other related integral in equation
(13.21) may also be found in standard tables. It will be seen that x_0 is a
scale factor, and may be thought of as some particular dimension of the
dome (say half the diameter at its base). Figure 13.13 is a plot of equation
(13.21); the curve shown differs from a quarter-circle of unit radius centred
near (0,1) by just over 4 per cent, the value shown in fig. 13.12 (for the range
x/x_0, $y/y_0 < 1$, cf. fig. 13.10).

A dome of any ratio rise/diameter can be constructed with the profile
of fig. 13.13, by taking a suitable segment of the curve. Figure 13.13 is the
quasi three-dimensional analogue of the catenary (Hooke's hanging chain),

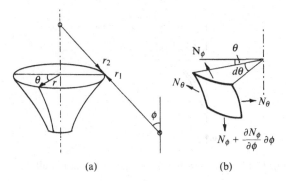

Fig. 13.14. (a) the geometry of the fan vault and (b) an element of the shell.

and the resulting dome of revolution will be free of hoop stress, carrying only meridional compressive stresses.

13.3 Fan vaults

The membrane analysis for the dome may be repeated for the fan vault. Figure 13.14 shows a complete fan-vault conoid, and solutions will be sought for which the hoop stress N_θ is zero, as before. If such a profile can be found, then the complete conoid may be cut in two by a vertical plane passing through the axis, and the resulting half-shell used as a model for the real fan vault.

The shell of fig. 13.14 has negative Gaussian curvature, and this introduces some minus signs into the membrane equations. Equations (13.13), (13.14) and (13.15) are replaced by

$$\frac{d}{d\phi}(rN_\phi) + r_1 N_\theta \cos\phi + wrr_1 \sin\phi = 0, \tag{13.22}$$

$$\frac{N_\phi}{r_1} - \frac{N_\theta}{r_2} = -w\cos\phi, \tag{13.23}$$

$$\left.\begin{array}{c} r = r_2 \sin\phi \\ \text{and} \quad r_1 \cos\phi = -\dfrac{dr}{d\phi} \end{array}\right\}. \tag{13.24}$$

Further, $\tan\phi = -dy/dx$ (see fig. 13.15) and the scale factor x_0 which arises in the integrations may be identified with the maximum radius of the fan vault at its crown, $y = 0$. At this condition the requirement $\phi = 0$ will not

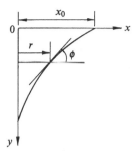

Fig. 13.15. Notation for the shape of the fan vault.

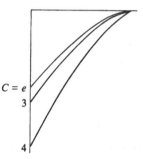

Fig. 13.16. Three 'perfect' shapes for the fan vault in which the form
follows the line of thrust.

be imposed, and the solution of equations (13.22), (13.23) and (13.24) with
$N_\theta = 0$ proceeds as for the dome, leading to

$$y = \frac{1}{2}Cx_0 \left[\int_{x/x_0}^1 e^{-t^2} dt - \frac{1}{C^2} \int_{x/x_0}^1 e^{t^2} dt \right]. \tag{13.25}$$

Equation (13.25) defines a family of curves generated by particular values of
the constant C. The angle ϕ is positive at $x = x_0$ for $C > e$, and this is the
lower bound of C of practical interest. In fig. 13.16 the profiles are sketched
of the generating curves for fan vaults given by $C = e$, 3 and 4.

Conditions at the edge of the fan are of interest. Figure 13.17 shows the
forces acting on complete half-conoids of the three profiles of fig. 13.16. The
horizontal thrust exerted by each vault has value wx_0^2, independent of the
value of the constant C. The upper edges of the vault require horizontal and
vertical forces as shown in order that overall equilibrium may be maintained.
The horizontal force will be generated by each fan leaning against the fan

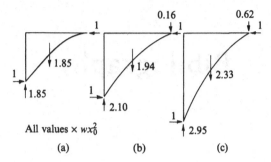

Fig. 13.17. Forces generated by the shapes of fig. 13.16.

on the opposite side of the church; the vertical forces can be provided by spandrel masonry on the centre line of the church, which often includes heavy masonry bosses.

A study was made by Anderson *et al.* (1965) of the best shape for a space vehicle entering thin planetary atmospheres. Effectively, a thin membrane was proposed as a sort of parachute, and the governing design criterion was that, to avoid wrinkling of the membrane, the stresses everywhere should be tensile. The problem is an almost exact 'Hookean' analogue of that of the masonry vault and, although the loading conditions are different, the parachute proposed is remarkably like a complete fan vault entering the atmosphere point first. At the widest part of the fan, the parachute edge must be reinforced by a stiffening ring; no solution exists for a free edge. This, again, is analogous to the compressive edge load required for the masonry fan vault.

Bibliography

CHAPTER 1

Ackermann, J.S. (1949). *Ars sine scientia nihil est,* Gothic theory of architecture at the Cathedral of Milan. *The Art Bulletin,* **31**, 84–111.

Bernoulli, D (1751). De vibrationibus et sono laminarum elasticarum. *Commentarii Academiae Scientiarum Petropolitanae,* **13** (1741–3), 105–20.

Bernoulli, James (1691). Specimen alterum calculi differentialis. *Acta eruditorum,* 13–23.

Bernoulli, James (1705). Véritable hypothèse de la résistance des solides, avec la démonstration de la courbure des corps qui font ressort. *Mémoires de l'Académie des Sciences,* 176–86.

British Standard 8110: Part I: 1985. *Structural use of concrete.*

Coulomb, C.A. (1773). Essai sur une application des règles *de maximis & minimis* à quelques problèmes de statique, relatifs à l'architecture. *Mémoires de Mathématique & de Physique, présentés à l'Académie Royale des Sciences par divers Savans, & lûs dans ses Assemblées,* **7**, 343–82.

Euler, L. (1744). *Methodus inveniendi lineas curvas maximi minimive proprietate gaudentes, sive solutio problematis isoperimetrici latissimo sensu accepti,* Lausanne and Geneva. Reprinted (1952) in *Leonhardi Euleri Opera Omnia,* 1st series, **24**, Bern.

Euler, L. (1757). Sur la force des colonnes. *Mémoires de l'Académie Royale des Sciences de Berlin,* **13**, 252–82.

Galileo Galilei (1638). *Dialogues concerning two new sciences* (trans. H. Crew and A. de Salvio, 1914, New York).

Heyman, J. (1973). Plastic design and limit state design. *The Structural Engineer,* **51**, 127–31.

Lagrange, J.L. (1770). Sur la figure des colonnes. *Miscellanea Taurinensia,* **5**, 123–66.

Mariotte, E. (1686). *Traité du mouvement des eaux,* Paris.

Navier, C.L.M.H. (1826). *Résumé des leçons données à l'Ecole des Ponts et Chaussées, sur l'application de la mécanique à l'établissement des constructions et des machines,* Paris. (3rd edn, with notes and appendices by B. de Saint-Venant, 1864, Paris.)

Parent, A. (1713). *Essais et recherches de mathématique et de physique,* 3 vols, Paris.

Steel Structures Research Committee, First Report (1931), Second Report (1934), Final Report (1936), HMSO, London.

CHAPTER 2

Baker, Sir John, and Heyman, J. (1969). *Plastic design of frames.* Vol. 1, *Fundamentals.* Cambridge University Press.

Heyman, J. (1982). *Elements of stress analysis,* Cambridge University Press.
Sokolnikoff, I.S. (1946, second edition 1956). *Mathematical theory of elasticity,* McGraw-Hill, New York.

CHAPTER 3

Beggs, G.E. (1927). The use of models in the solution of indeterminate structures. *Journal of the Franklin Institute,* **203,** 375.
Betti, E. (1872). Teorema generale intorno alle deformazioni che fanno equilibrio a forze che agiscono soltanto alle superficie. *Il Nuovo Cimento,* ser.2, **7,** 87 and **8,** 97.
Charlton, T.M. (1966). *Model analysis of plane structures,* Pergamon Press, Oxford.
Charlton, T.M. (1982). *A history of the theory of structures in the nineteenth century,* Cambridge University Press.
Maxwell, J.C. (1864). On the calculation of the equilibrium and stiffness of frames. *London, Edinburgh and Dublin Philosophical Magazine,* ser. 4, **27,** 294.
Melchers, R.E. (1980). Service load deflexions in plastic structural design. *Proc. Instn civ. Engrs,* **69,** 157.
Müller-Breslau, H.F.B. (1883). Zur Theorie der versteifung labiler flexibiler Bogenträger. *Zeitschrift für Bauwesen,* **33,** 312.
Müller-Breslau, H.F.B. (1884). Einflusslinien für kontinuerliche Träger mit drei Stutzpunkten. *Zeitschrift des Architekten und Ingenieur-Vereins zu Hannover,* **30,** col. 278.

CHAPTER 4

Heyman, J (1993). The roof of the monks' dormitory, Durham; in *Engineering a Cathedral,* Thomas Telford, London, pp. 169–79.

CHAPTER 5

Case, J. (1925). *Strength of materials,* Arnold, London.
Clebsch, A. (1862). *Theorie der Elasticität fester Körper,* Leipzig. Translated by Saint-Venant and Flamant, with notes by Saint-Venant (1883), *Théorie de l'élasticité des corps solides,* Paris.
Lowe, P.G. (1971). *Classical theory of structures,* Cambridge University Press.
Macaulay, W.H. (1919). Note on the deflection of beams ..., *Messenger of Mathematics,* **48,** 129.
Wittrick, W.H. (1965). A generalization of Macaulay's method with applications in structural mechanics, *American Institute of Aeronautics and Astronautics Journal,* **3,** 326.

CHAPTER 6

Bernoulli, James (1694). Curvatura laminae elasticae, *Acta eruditorum Lipsiae.*
Bernoulli, James (1695). Explicationes, annotationes et additiones, *Acta eruditorum Lipsiae.* Both papers reprinted (1744) in *Opera* (2 vols), Geneva.

Euler, L. (1744). *Methodus inveniendi lineas curvas maximi minimive proprietate gaudentes, sive solutio problematis isoperimetrici latissimo sensu accepti*, Lausanne and Geneva. Reprinted (1952) in *Leonhardi Euleri Opera Omnia*, 1st series, **24**, Bern.

Euler, L. (1757). Sur la force des Colonnes. *Mémoires de l'Académie Royale des Sciences et Belles Lettres*, **13**, 252, Berlin.

Truesdell, C. (1960). *The rational mechanics of flexible or elastic bodies (1638–1788)*, *Leonhardi Euleri Opera Omnia*, 2nd series, **11(2)**, Zürich.

CHAPTER 7

Baker, Sir John and Heyman, J. (1969). *Plastic design of frames*, Vol. 1, *Fundamentals*, Cambridge University Press.

Heyman, J. (1971). *Plastic design of frames*, Vol. 2, *Applications*, Cambridge University Press.

Heyman, J. (1975). Overcomplete mechanisms of plastic collapse. *Journal of Optimization Theory and Applications*, **15**, 27.

CHAPTER 8

Drucker, D.C. and Shield, R.T. (1957). Bounds on minimum weight design. *Quarterly of Applied Mathematics*, **15**, 269.

Heyman, J. (1971). *Plastic design of frames*, Vol. 2, *Applications*. Cambridge University Press.

CHAPTER 9

Heyman, J. (1959). Inverse design of beams and grillages. *Proc. Instn civ. Engrs*, **13**, 239–52.

Nervi, P.L. 1956. Concrete and structural form. *The Structural Engineer*, **34**, 155.

CHAPTER 10

Heyman, J. (1972). The significance of shakedown loading, *9th Congress, International Association for Bridge and Structural Engineering*, Preliminary Report, Amsterdam.

Ogle, M.H. (1964). Shakedown of steel frames. PhD thesis (Cambridge University).

CHAPTER 11

Coulomb, C.A. (1773) Essai sur une application des règles de maximis & minimis à quelques problèmes de statique, relatifs à l'architecture, *Mémoires de Mathématique & de Physique, présentés à l'Académie Royale des Sciences par divers Savans, & lûs dans ses Assemblées*, **7**, 343–82.

Emerson, W. (1754; new edition 1825). *The principles of mechanics...*, London.

Heyman, J. (1972). *Coulomb's memoir on statics*, Cambridge University Press.

Mariotte, E. (1686). *Traité du mouvement des eaux*, Paris.

Navier, C.L.M.H. (1826). *Résumé des leçons données à l'Ecole des Ponts et Chaussées, sur l'application de la mécanique à l'établissement des constructions et des machines*, Paris. (3rd edn, with notes and appendices by B. de Saint-Venant, 1864, Paris.)

Parent, A. (1713). *Essais et recherches de mathématique et de physique*, 3 vols, Paris.

'T.T.' (1816). Rules for ascertaining the Strength of Materials, *The Philosophical Magazine and Journal*, **47**, 21.

Tredgold, T. (1824). *Practical essay on the strength of cast iron*, second edition, London.

CHAPTER 12

Brown, E.H. (1967). Plastic asymmetric bending of beams. *International Journal of Mechanical Sciences*, **9**, 77–82.

Ewing, J.A. (1899). *The strength of materials*. Cambridge University Press.

Heyman, J. (1968). The simple plastic bending of beams. *Proc. Instn civ. Engrs*, **41**, 751–9.

CHAPTER 13

Anderson M.S., Robinson C.J., Bush H.G. and Fralich R.W. (1965). A tension shell structure for application to entry vehicles. National Aeronautics and Space Administration Technical Note NASA TN D–2675, Washington DC.

Heyman, J. (1967). On shell solutions for masonry domes. *International Journal of Solids and Structures*, **3**, 227–41.

Heyman, J. (1967). Spires and fan vaults. *International Journal of Solids and Structures*, **3**, 243–57.

Heyman, J. (1977). *Equilibrium of shell structures*, Oxford University Press.

Heyman, J. (1988). Poleni's problem. *Proc. Instn civ. Engrs*, **84**, 737.

Heyman, J. (1992). Leaning towers. *Meccanica*, **27**, 153–159.

Hooke, R. (1676). *A description of helioscopes, and some other instruments*. London.

Jahnke, E. and Emde, F. (1945). *Tables of functions*, 4th edn. Dover Publications, New York.

Poleni, G. (1748). *Memorie istoriche della gran cupola del Tempio Vaticano*. Padova.

Name index

Subject index